A Field Guide to
THE NORTON BOOK OF
NATURE WRITING

College Edition

 LILACE MELLIN GUIGNARD

University of Nevada, Reno

W·W·NORTON & COMPANY·NEW YORK·LONDON

For information about permission to reproduce selections from this book, write
to Permissions, W. W. Norton & Company, Inc., 500 Fifth Avenue, New York,
NY 10110

The text of this book is composed in Electra with the display set in
Bernhard Roman.
Composition by Tom Ernst.
Manufacturing by Phoenix Color.

ISBN 0-393-97815-X (pbk.)

W. W. Norton & Company, Inc., 500 Fifth Avenue, New York, NY 10110
www.wwnorton.com

W. W. Norton & Company Ltd., 15 Carlisle Street, London W1D 3BS

2 3 4 5 6 7 8 9

CONTENTS

6

THEMATIC
TABLE OF CONTENTS

ENCOUNTERS WITH AND AMONG CREATURES

DEFINING SPACES:
EDENS, EDGES, WILDERNESS, AND URBAN NATURE

WORKING (WITH) THE LAND AND ANIMALS

DWELLING IN PLACE

KNOWING NATURE, KNOWING OURSELVES

NATURAL PHENOMENA

15

PREFACE

THE GOAL OF THIS GUIDE is to aid students and teachers in thinking about and responding to the readings in the *Norton Book of Nature Writing*. Those of us who study nature writing have various reasons for doing so. In literary and rhetorical studies, the study of nature writing reveals some of the ways that language and literature both reflect and shape our relationships to the nonhuman world. It helps us to see how language comes *from* the earth and how our attempts to understand and explain natural phenomena result in some of our most fundamental — and powerful — cultural metaphors.

But why study literature in environmental studies — or history, or philosophy, or geography? Because it makes connections between students, the classroom, and the world and therefore makes for good pedagogy. Good narratives convey way more than just "facts"; they say something about the significance of those facts. And the facts themselves take on larger meaning when seen alongside the literature of the natural world.

This guide can be used in many ways. Like most field guides, it groups materials by classifications, starting with a thematic table of contents that groups the selections by key themes — encounters with and among creatures, writing by explorers, and so on. Within each theme, selections are listed chronologically, because I feel that some understanding of the historical context is imperative to a reader's understanding and appreciation. There are many essays that could fall under more than one heading, of course. When this was the case, I asked myself what was the strongest message of the piece, and whether it would serve and be served by the others in that category. The fact that many nature writers reflect on death, or on the animals with which we share the land, as well as on scientific facts and theories, is what makes this literature so compelling and inventive.

This guide also includes study questions for each selection in the

Norton Book of Nature Writing. The questions themselves follow a loose pattern. The first focuses the reader on details of the essay, either content or form. The second suggests and asks for comparisons between writers and texts. The third is usually multi-layered and meant as a discussion or writing prompt, often using the essay as a jump-off point for reflecting on and offering one's own opinion concerning a topic the writer raises. Occasionally the first question will have a specific, non-negotiable answer, but in most cases the questions may elicit many varying responses. Indeed, many of these essays critique the idea that there can be an objective point of view or an absolute truth. It is my hope that these questions lead to discussions in which students and teachers together explore the complexities of human interaction with nonhuman nature, and that classrooms allow for multiple points of view.

In the introduction to the First Edition, Elder and Finch say that "all literature asks a single question: how shall we live?" I agree—and I composed this guide to approach the texts from a variety of perspectives, but always with that question at the core. As Scott Russell Sanders says in his essay "Buckeye," "nothing in my education prepared me to love a piece of the earth." We can begin to change that. We must. And if John Fowles is right that "achieving a relationship with nature is both a science and an art," students of all disciplines are needed to make this change happen. The issues far exceed any one discipline's perspective.

This is why I study and teach nature writing, and I hope that this field guide will give other readers similar motivation.

Lilace Mellin Guignard
University of Nevada, Reno

GILBERT WHITE (1720–1793)

From THE NATURAL HISTORY AND ANTIQUITIES OF SELBORNE (pp. 33–50)

1. According to the the headnote, Gilbert White "typified the ideal of amateur science." How does White use keen observation and comparisons between species (including humans) to theorize about non-human creatures?

2. When discussing the phenomenon of echoes and the mythic/poetic way Echo has been personified, White says, "Nor need the gravest man be ashamed to appear taken with such a phenomenon, since it may become the subject of philosophical or mathematical inquiries." What are the differences between the "beautiful fiction[s]" about Echo and other natural occurrences and the facts and theories philosophers and mathematicians derive from their observations? What are the goals of each mode of investigation, and how does White combine them?

3. Supposition is a manner of questioning, of saying "Suppose. . . ." How does White use supposition in his scientific process? When discussing the swifts, White asks—but does not answer—many questions. Are you accustomed to being given answers in scientific writing? How would White's writing be different if he'd had some of the field guides and other resources available to us now? Do you think you would like it more, or less? Why?

HECTOR ST. JOHN DE CRÈVECOEUR (1735–1813)

From LETTERS FROM AN AMERICAN FARMER and SKETCHES OF EIGHTEENTH CENTURY AMERICA (pp. 51–63)

1. Crèvecoeur refers to himself as the law in his barnyard, "a bridle and check to prevent the strong and greedy from oppressing the timid and weak." How does he do this? Are there times when his interventions

do not seem just or fair, or times when you think he should have intervened but did not?

2. Crèvecoeur is judgmental about other humans' interventions. How do you think he would feel about Gilbert White's untiling the eaves of a house and handling the swifts or about White's neighbor throwing the hawk that killed his chickens into the chicken pen to be tortured? What does Crèvecoeur believe about humans' responsibilities toward other creatures? Cite examples from Crèvecoeur's writing to support your opinion.

3. Crèvecoeur characterizes the contest between the black snake and water snake and the king-bird's attacks on other birds as "uncommon and beautiful," not "an unpleasant sight to behold." How do you feel about scenes in which creatures die or are hurt? Do you find yourself cheering for one creature over the other? What qualities capture your empathy? Do you feel more attached to those creatures you find beautiful? To ones with which you feel you have something in common?

WILLIAM BARTRAM (1739–1823)

From TRAVELS THROUGH NORTH & SOUTH CAROLINA, GEORGIA, EAST & WEST FLORIDA . . . (pp. 64–76)

1. Why is it important for naturalists to name plants and creatures? What is the importance of names to writers? Did the names help or hurt your understanding of the selection?

2. Linnaeus (1707–1778) was a Swedish botanist who developed the two-word system for classifying living things: the first word is the genus and the second is the species. Bartram added to the Linnaean system, collecting and naming new specimens. How do his tone and language reflect the natural history knowledge of his time and his desire to find patterns in what he saw?

3. Using a field guide or encyclopedia, do a little research to find out what we know now about some of the creatures that Bartram and other eighteenth-century writers observed and made claims about. Compare the language of these "scientific" sources to that in Bartram's writing. Does anything you discover make you now doubt his credibility?

ALEXANDER WILSON (1766–1813)

From AMERICAN ORNITHOLOGY; OR,
THE NATURAL HISTORY OF THE BIRDS OF
THE UNITED STATES (pp. 77–81)

1. Wilson compiled one of the early multivolume field guides about birds. What previous notions about the ivory-billed woodpecker does he discredit and how?

2. Wilson capitalizes the word *Nature*, as does Crèvecoeur; Bartram and White do not. Compare the ideals they hold about the relationship between Man, Nature, and God, and consider how these ideals are reflected in their writing style. Does style have any connection with the writers' ideals? Does a more literary style align with a certain outlook, and a more scientific style with another? Cite examples from their texts to support your answer.

3. You are a naturalist. Describe a creature as if you were writing a field guide entry like Wilson's. It can be the field guide to anything animate—dogs, little brothers. (It does not have to be another species, but it does have to be something that you can directly observe or with which you are extremely familiar.) Start with a general description; then add details that reveal characteristics, habitat, range/movement, and behavior. Be sure to point out common misconceptions and your disagreements with common perceptions, supporting your views with specific examples. Record unusual sightings, and describe at least one significant interaction you have had with "the creature." Include its home-building techniques; its common names and its scientific or specific name; and its song, call, or distinctive noises. Then give a precise physical description, including measurements, color, anatomy, and any known variations. Do any of these traits correspond to age, season, or habitat? What does it eat and how do you know? If using a member of your own species, ignore anything you know about him or her that you learned through language. What skills did you use in describing your own species as a naturalist?

JOHN LEONARD KNAPP (1767–1845)

From THE JOURNAL OF A NATURALIST (pp. 82–85)

1. In the headnote, the Linnaean model is considered a way for gentle-men to "organize days of rambling and observing out of doors." The Linnaean system of classifying all life forms, imposing a structure on the natural world, conveys a system of civilized behavior. In what other ways does Knapp demonstrate that he is a gentleman con-cerned with civil behavior?

2. The term *gentleman* may denote a male belonging to a privileged socioeconomic class. Where do Knapp's prejudices against others of his species show up in his writing? Compare him to other nature writ-ers you've read so far. Which writers used nature writing to "organize days of rambling"? Which wrote of nature in connection with their profession? Which made their observations while working with the land? How might the profession and/or social position of a nature writer affect his perceptions and style of writing?

3. Knapp, like other nature writers, is concerned with the ways cultural prejudices often ignore science and reason, resulting in unnecessary cruelty toward certain species. He also notes the power of language to preserve prejudices long after we know better, such as by the use of words like *vermin* and *shrew*. Do you have cultural prejudices toward the natural world that have not been justified by your personal experience? In current environmental debates, do you recognize cul-tural prejudices in some arguments? Why are pit bulls, coyotes, star-lings, squirrels, spiders, snakes, and other species often called "bad" or "pests"? Is it wrong to use antibacterial soap? Is it cruel to kill weeds? Choose one of these issues or pick another one with which you're familiar and show how popular opinion is based on prejudice rather than on scientific knowledge. Which human values determine the image, positive or negative, that your subject has acquired?

DAVID THOMPSON (1770–1857)

From DAVID THOMPSON'S NARRATIVE OF HIS EXPLORATIONS IN WESTERN AMERICA, 1784–1812 (pp. 86–89)

1. Thompson declares that he is describing traits "which naturalists do not, or have not noticed." What traits is he referring to, and why do you think naturalists of his time did not notice or mention them?

2. What attracts Thompson to the animals he describes? Compare his literary and physical treatment of these animals to that of other nature writers like White, Crèvecoeur, Wilson, Bartram, and Waterton. Some writers portray animals as having individuality, and others see all members of a species as the same. Discuss the writers' language as it reflects and enforces their views of animals.

3. In relaying his experience of the Aurora Borealis, Thompson contrasts reason and sense. "Sense" refers here to sensory perception, such as seeing and hearing. When the wolverines take the axes, some sense that fairies or ghosts had to have done it because no humans were around. Think of an experience you have had, or that you have heard about, in which the cause is argued about in terms of reason and senses. For example, someone says, "That couldn't happen," and another replies, "But I *saw* it." Did your parents ever disbelieve you as a child? Do you believe stories of Bigfoot sightings? In deciding what to believe, do you trust your reason or your senses more?

DOROTHY WORDSWORTH (1771–1855)

From JOURNALS OF DOROTHY WORDSWORTH (pp. 90–95)

1. How does Wordsworth achieve "a lyrical immediacy that makes a reader see what she has seen, feel what she has felt"? Make a list of passages in which this seems true, and identify the rhetorical devices she uses to achieve this effect.

2. Compare Wordsworth's style of lyrical immediacy with Bartram's excited narrative. Wordsworth focuses more on images and moods, while Bartram concentrates on story and suspense. Which approach do you prefer? Why? Each style reveals directly or indirectly a great

deal about the writer. What do you learn about each from their essays?

3. In these selections from her journals, Wordsworth demonstrates a heightened awareness of light and an agility at conveying how light affects a landscape as perceived by the walker. This is one way she goes beyond identifying and reporting on the behavior of creatures to describe how scenes, walks, days *felt* to her. Think about a time when you were outdoors, perhaps watching a sunset or walking in rain, and you felt a specific mood in the atmosphere. Describe the scene in a way that conveys the mood as well as the reality. Did the mood of the scene correspond with your mood? Did it affect your mood? Either state directly what you were thinking before and during the experience, or use images and metaphors to suggest your mood at the time.

MERIWETHER LEWIS (1774–1809)

From THE JOURNALS OF LEWIS AND CLARK (pp. 95–104)

1. Lewis's observations emphasize two senses not commonly used in other essays. What are they? Give examples from the text.

2. Such a great and early exploration carried the terrific responsibility of describing new animals and lands in ways that would be easily understood by those back east. Lewis often describes things in terms of value: referring to an animal's or place's value to humans and other creatures. He also makes distinctions between humans and animals, civilized people and Indians, civilized places and natural places. What does this focus on distinctions tell you about what the explorers valued most? Compare Lewis and his party to Thompson, Waterton, and Catlin. How did the goals of these explorers shape their values and subsequently their descriptions?

3. Lewis describes the buffalo and other mammals as observing the white explorers seemingly with as much interest as the humans observe the animals. Are you comfortable imagining events from the animal's view? How does the implication that animals have curiosities and emotions similar to our own complicate your experience of these texts?

CHARLES WATERTON (1782–1865)

From WANDERINGS IN SOUTH AMERICA,
THE NORTH-WEST OF THE UNITED STATES,
AND THE ANTILLES (pp. 104–16)

1. Environmental historian Donald Worster notes that the collecting and categorizing of Linnaeus and his followers often followed an " 'imperial' view of nature" owing to Western civilization's belief that humans have dominion over the natural world and that they need to catch, cage, and sometimes kill creatures in order to study them. How is this utilitarian view of nature manifest in Waterton's writing?

2. Gilbert White writes, "[M]y little intelligence is confined to the narrow sphere of my own observations at home." In contrast, Waterton writes about exploring unfamiliar lands. Like Bartram, he wrote for general readers. Like Lewis, he felt charged with a responsibility to collect information for those back home. What other comparisons in purpose and style can you make among Waterton, Bartram, and Lewis, especially as compared with White, Crèvecoeur, and Wordsworth, who roamed repeatedly over familiar terrain?

3. Waterton clearly believes that animals exist as a natural resource for human use, like oil, coal, or water. How do you feel about this belief? Do humans have any moral obligations to other creatures? How and where do you draw the line between what is okay and not okay to do to other creatures?

JOHN JAMES AUDUBON (1785–1851)

From ORNITHOLOGICAL BIOGRAPHY (pp. 117–22)

1. Audubon describes three scenes of predation. Define the predators and the prey in each scene. Do you see similarities in the predators' intent and in how the characters are represented?

2. Audubon believed that God created a balanced natural economy with little or no waste. How does the sheepherder's reaction to the wolves fit within the balance of nature? Compare the sheepherder's intervention to Crèvecoeur's interventions and views. Look especially

at Audubon's first paragraph. Do you think he sees humans as having an unfair or unjustified prejudice against wolves?

3. Write about a situation in which you felt like either predator or prey. Explain why you felt that way.

JOHN CLARE (1793–1864)

THE NATURAL WORLD and from THE NATURAL HISTORY PROSE WRITINGS OF JOHN CLARE (pp. 122–28)

1. Clare is a Romantic. Why does he look to nature if not to measure and catalogue it? How does he feel about taking specimens?

2. In his letter of February 7, 1825, Clare quotes Chaucer on nightingales and later expounds upon the relationship between images in poetry and images in nature. He distinguishes between the "man of taste" who can look on the celandine flower and recall William Wordsworth's poem, and the "country clown" who knows nothing of poetic pleasures. Why does knowing the poetic associations deepen Clare's experience of the flower? Compare Clare's use of verse in these passages to White's references to Ovid, Virgil, and others.

3. Nature writers often address the relationship between book learning and learning from experience. At first, Clare seems to elevate poetry describing natural objects over the actual experience of those objects. Later in his letter, however, he shows the superiority of direct sensory perception, when he points out that any "country clown" knows nightingales sing during the day because he has *heard* them. Still, because we cannot see, hear, taste, smell, and touch everything firsthand, we rely on accounts such as those of naturalists and poets to communicate the experience of places and creatures through language. Do you think words on the page can convey experiences, or is reading the opposite of experiencing? What responsibility do writers have to relay events and describe natural objects accurately?

GEORGE CATLIN (1796–1872)

From LETTERS AND NOTES ON THE MANNERS, CUSTOMS, AND CONDITIONS OF THE NORTH AMERICAN INDIANS . . . (pp. 129–40)

1. Even though Catlin refers to paintings or plates we cannot see, he describes scenes in great detail—especially the buffalo's wallowing in self-made puddles. Catlin also observes how these buffalo-made sinks later become odd circles of flowers and grasses in the middle of the prairie. How might Catlin's training as a painter have helped him make such connections between phenomena and describe them in such detail?

2. Catlin highlights the practices that he believes will lead rapidly to the extinction of the buffalo. Clare and White mention diminishing populations of birds, and John Knapp discusses the elimination of the marten's habitat by intruding human populations, stating that "wild creatures are yearly decreasing, and very many that once existed have even become extinct." Compare these writers' attitudes toward species extinction. How does each investigate and explain the causes?

3. As Euro-American settlers moved westward, the way humans interacted with the herds changed greatly. In the 1820s, there were more buffalo (American bison) than people in North America. In 1830, buffalo destruction was advocated by the U.S. government as a means of subduing hostile Indians through starvation. The buffalo population dwindled to about eight hundred in 1895, after which rescue and restoration efforts brought the numbers up to approximately thirty thousand by 1996 (all within national parks and refuges; www.amnh.org). Using Catlin's descriptions and your own knowledge of Native American history, discuss what the buffalo and the Native Americans have in common. How did the Indians' behavior toward the buffalo change when white settlers moved west?

RALPH WALDO EMERSON (1803–1882)

From NATURE and THE JOURNALS OF
RALPH WALDO EMERSON (pp. 140-51)

1. Emerson was one of the first advocates of Transcendentalism, the major strain of Romanticism in America. To communicate his abstract reasoning and ideas, Emerson often divides a subject into two categories in order to emphasize the presence and absence of a certain trait. What are the dualisms Emerson establishes? What distinction does each emphasize?

2. Though Emerson did experience nature directly, he approached it less as a scientist and more as a poet-philosopher. "The woods were too full of mosquitoes," he says, "to offer hospitality to the muse." Like Clare, Emerson looks to nature for inspiration. Unlike Clare, Emerson insists on an emblematic relationship between the outer world and the inner. Chiefly through metaphor, Emerson leads the reader to deeper understanding—not just appreciation. Explore further the ways in which these two writers view the relationship between art and nature.

3. Emerson focuses on the spiritual aspects of an individual's relationship to nature. As a Transcendentalist, he disdains intermediaries like priests, church rituals, or books (such as the Bible). He interprets moral lessons directly from nature, believing that to experience nature is to behold God "face to face." Because of his spiritual orientation, Emerson's questions and answers differ from those of nature writers who take a more scientific approach. What do you think of Emerson's inquiry? What questions do *you* have concerning nature's purpose? Are they spiritual, philosophical, scientific, or all three?

CHARLES DARWIN (1809–1882)

From VOYAGE OF *H.M.S. BEAGLE*, ON THE ORIGIN OF
SPECIES, and THE DESCENT OF MAN (pp. 151–63)

1. In his accounts of the *Beagle* voyage, Darwin is aware of how his impressions of "primitive" places and creatures are shaped by their contrast with the "civilized" places to which he is accustomed. By the early nineteenth century, Western civilization was rapidly spreading

across the globe. Identify passages in which Darwin alludes to this phenomenon. How does his language reveal his attitude toward the places and people he encounters?

2. Emerson wrote, "[W]henever a true theory appears, it will be its own evidence." Darwin's theory is widely accepted today, but when he first published *On the Origin of Species* and *The Descent of Man*, the mere idea of evolution was so radical that he consciously employed rhetorical techniques to make it easier to accept, including the statement that his view "ennobled" all beings. Emerson made strong claims too. Compare the way each of these men investigated nature and convinced readers of their theories concerning nature.

3. In *The Descent of Man*, Darwin's ideas on regulating marriage and procreation to self-select for certain qualities are informed both by scientific knowledge and by his imperialist interests. With all the genetic engineering currently underway, as well as research on the human genome and cloning, what do you think are the risks and/or benefits of this type of knowledge? Now that we know much more about "the laws of inheritance," as Darwin says, can we partially realize his Utopia?

SUSAN FENIMORE COOPER (1813–1894)

From RURAL HOURS (pp. 163–68)

1. Cooper writes of her home in Cooperstown, New York, and of the changes humans have made there in the last fifty years. What is her attitude? Cooper was the first American woman to publish a book of nature writing. What do you think her goal was in writing this book?

2. In this selection, Cooper is concerned mainly with the landscape and the changes made in it by agricultural settlement. The only creatures she refers to are humans and cattle. Compare her account with Catlin's record of the effects of settlement on the western frontier. What might account for the tonal differences between Cooper's and Catlin's writing?

3. The uses animals and humans make of a place can be revealing, such as Cooper's cattle terraces on the hillsides. Write a few pages describing a place you know well, documenting signs of use, disuse, settlement, and abandonment. It can be a field you played in, a backyard, or even a house you lived in for some time. How can you tell who

used the place? Can you tell when and what for? What clues are there about who uses it now? Let your feelings (good or bad) for the place come through in your descriptions.

HENRY DAVID THOREAU (1817–1862)

From A WEEK ON THE CONCORD AND MERRIMACK RIVERS, WALDEN, WALKING, THE MAINE WOODS, *and* JOURNALS (pp. 168–220)

1. Another prominent Transcendentalist and a student of Emerson's, Thoreau begins "Walking" by saying he wants to "regard man as an inhabitant, or a part and parcel of Nature." Drawing from any of Thoreau's writings, identify passages that demonstrate this view. How does Thoreau represent the relationship between humans and animals? How does he use metaphors to reinforce this perspective?

2. In the examples of polished, published works by both Emerson and Thoreau, as well as excerpts from their journals, determine the way each uses his journal as a "literary laboratory." What differences and similarities do you find between the narrators in the essays and in the styles and attitudes in the journals?

3. Thoreau is aware of himself as "both observer and participant." Perhaps this awareness captures humankind's strange position in regard to the rest of nature—our highly evolved consciousness makes us both *of* it and *aware of* it. John Elder explains how

   > Thoreau projects a powerful image of the interaction of mind and world. Left foot/right foot, the walker moves through the landscape. Right brain/left brain, sensation and reflection flicker into the complex wholeness of human response.
   >
   > —from the Introduction to *Nature/Walking*

   What is the relationship between the interior and exterior worlds? Are body and spirit separate entities, the Me and the Not Me, as Emerson says? How can Thoreau not be where his body is?

CHARLES KINGSLEY (1819–1875)

From GLAUCUS; OR, THE WONDERS OF THE SHORE (pp. 220–23)

1. Kingsley has a distinctively animated narrative voice. How does he directly and informally address you, the reader, drawing you into the passages as if into conversation with him?

2. Darwin, the scientist, referred to variability as a natural law that, once the Creator "breathed . . . into a few forms or into one," caused, through "the indirect and direct action of the conditions of life," all the multiple life forms to evolve. Kingsley, a priest, explains that the "prodigal variety" exists because "God has created all things for Himself, and rejoices in His own handiwork." Compare Darwin's and Kingsley's theories and the possible consequences of each one.

3. *Anthropocentric*, which means "human-centered," refers to the view that the world was made for humans and that our species is the climax of creation or evolution. The "imperial ecologists" were anthropocentric, as are those who believe that God gave humans dominion over the beasts and land. But Darwin is an example of a scientist who hoped his theory of evolution would make humans realize that they are kin to the animals. And Kingsley is an example of a priest who preached that man is not "the centre and the object of [creation's] existence," that nature does not belong to humans "but to One greater." Read and analyze several nature writers, asking yourself what humans' relationship to nature is or should be. Then choose a writer whose views are close to yours, and challenge other writers with whose views you disagree. Use current events and personal experiences to support your views.

WALT WHITMAN (1819–1892)

From SPECIMEN DAYS AND COLLECT (pp. 223–29)

1. The direct experience of nature was important to Whitman, as it was to Emerson and Thoreau. Whitman repeatedly asserts that he is writing outside, in nature. Identify sections where Whitman calls the reader's attention to where he is, placing the reader in a specific place

and time. How does he achieve a greater sense of immediacy and intimacy by doing this?

2. Like Thoreau and the Transcendentalists, Whitman wished not to know too much and not to name everything. He equates a "free margin" with "ignorance," similar to Thoreau's belief that our "conceit that we know something . . . robs us of the advantage of our actual ignorance." Thoreau says "the highest that we can attain to is not Knowledge, but Sympathy with Intelligence." Where and how does Whitman demonstrate his feeling for other, nonhuman intelligences?

3. In "Seashore Fancies," Whitman tells how the shoreline has always suggested to him a "dividing line, contact, junction, the solid marrying the liquid." Later, Whitman speaks of his fusing with seas and shore, of them absorbing and understanding each other. This claim might seem strange, but the philosopher Merleau-Ponty's theory of experience supports it. He reasoned that perception is always interactive between the perceiver and the perceived: just as when your hand brushes against your knee and both your hand and knee feel the other, so a tree feels you when you feel the tree, and the robin perceives you as you perceive it. How would your life change if you believed that all things could feel and perceive you as well? What do you think such interaction means to Whitman?

JOHN WESLEY POWELL (1834–1902)

From EXPLORATION OF THE COLORADO RIVER OF THE WEST AND ITS TRIBUTARIES (pp. 230–36)

1. In Powell's first entry, he speaks about what is unknown. Is he comfortable with the situation? Exhilarated? Frightened? How can you tell?

2. Compare Powell's piece about descending the Colorado River with Thoreau's about traveling the Concord and Merrimack. How do their tones and descriptions differ? What accounts for these differences?

3. Powell and his party must constantly adjust to being in and out of control. Although Powell admits their lack of knowledge and control over nature, he gives detailed descriptions of men lining the boats down dangerous rapids, showing that they exert control in order to reduce the risks. Later, when they decide that portaging is impossible, "there is no hesitation." This fluid interplay between exercising

and relinquishing control is one reason Powell's adventure is so compelling. Write about a time when you had to struggle with taking or losing control over some nonhuman force. How did you decide what to do, and were you successful?

SAMUEL CLEMENS
(MARK TWAIN)
(1835–1910)

From LIFE ON THE MISSISSIPPI (pp. 236–38)

1. In this selection, Clemens boasts about his (and other steamboat captains') skill at reading the river. He is a pragmatist, a realist, but in some ways he regrets shedding his Romantic vision. How does he communicate his regret?

2. Kingsley wrote, "What a waste of power, on any utilitarian theory of nature." Clemens describes how his appreciation of the Mississippi River has become utilitarian rather than aesthetic or intuitive: "All the value any feature of it had for me now was the amount of usefulness it could furnish toward compassing the safe piloting of a steamboat." Apply Kingsley's critique to Clemens. How would Kingsley reply to Clemens's last question in the essay?

3. The literary shift from American Romanticism to Realism involved a shift in the approach to nature from wonder and a search for meaning to a pragmatic approach and a search for practical—not symbolic—truth. In Clemens's piece, we see his desire to have both approaches, though he does not believe that is possible. Do you think we can study and use nature while retaining a sense of awe toward it, a connection to faith and instincts?

CELIA THAXTER (1835–1894)

From AN ISLAND GARDEN (pp. 239–43)

1. Thaxter crafts her descriptions to create certain moods. Twice in this selection she makes an abrupt transition in mood from threatening or despairing to life-giving and joyous—once with the weather and once

with the hummingbird. Pinpoint these transitions. How does their abruptness affect your reading?

2. Thaxter tends her garden as Crèvecoeur tended his fields. Compare the way these two describe their interaction with the land and its inhabitants. Is one more pragmatic? Is one more Romantic?

3. Have you ever grown plants or nursed an animal to health? Raising a living thing may cause us to identify with it, so that we take it personally if the weather or other creatures harm it. Some argue that we should never interfere with natural forces. Others argue that we ourselves are animals and natural, so our intervention is also natural. How far should we go in our gardens, for example, to protect the plants we put there? Should we drive stakes and trellises in for support? Should we use pesticides and herbicides? Should Thaxter have left the hummingbird alone?

JOHN BURROUGHS (1837–1921)

IN MAMMOTH CAVE (pp. 244–50)

1. In the first paragraph, Burroughs tells how "the blind seem as much impressed by [Mammoth Cave] as those who have their sight." He proposes that beauty is not just visual but is something all senses experience. Look closely at his descriptions. How does Burroughs decide what is beautiful?

2. Burroughs relates the guide's stories of women's responses to the cave, saying "timid, highly imaginative people, especially women, are quite sure to have a sense of fear in this strange underground world." Thoreau, in "Walking," says, "[W]omankind . . . are confined to the house still more than men." How might nineteenth-century women's indoor lives have affected their relationship to nature? Are Burroughs and Thoreau judging women unfairly or are their observations accurate relative to their eras?

3. Burroughs's lyrical description of "the behavior of the cool air which welled up out of the mouth of the cave" draws attention to air as part of our environment. Perhaps this is an element of the landscape that the blind are more aware of than the sighted. Traditionally, any natural resource we take for granted is eventually abused. What do we as a species do that affects air? Do you think our actions would be different if we could see air? Why, or why not?

JOHN MUIR (1838–1914)

A Wind-Storm in the Forests *and*
The Water-Ouzel (pp. 250–68)

1. Muir depicts the water-ouzel in ways that communicate his personal attachment. What traits especially draw him to this bird? Why?

2. Emerson, Thoreau, and Muir are three of the patriarchs of American nature writing. We can see a successive increase in the extent to which each one physically interacts with nature. The activity and the elements are more extreme in Muir's work. He was not just a walker; he was a mountaineer exploring steep terrain, camping his way through pathless wilderness without hi-tech gear. How does Muir's increased physicality shape his literary style? How do his actions shape your response? Do they affect his credibility for better or worse?

3. Though Muir did eventually marry, he was a quintessential loner, like many other explorers. In "The Water-Ouzel," the bird is said to never get lonely and cheers Muir in his "lonely wanderings." Since both are "usually found singly," Muir looks to the ouzel (as well as trees and other nature) for companionship. Do you look to certain places or creatures for companionship? Did you, when you were a child? Could a tree, creek, or bird be better company at times than other humans?

W. H. HUDSON (1841–1922)

My Friend the Pig *and*
from Idle Days in Patagonia (pp. 269–75)

1. Hudson has an unrelenting sense of humor and word play. Who and what does he poke fun at? What is the effect?

2. Hudson discusses Thoreau, with whom he has much in common. Yet Hudson's humor, ridicule of polite society, and longing for a "vanished harmony between organism and environment" are more intense than Thoreau's. Not only was Hudson writing later than Thoreau, but England had lost its wilderness and became industrialized before the United States. How does Hudson's historical context, in contrast to Thoreau's, influence him and his writing?

3. Hudson uses humor to make the reader face the facts that pigs (and other creatures raised for food) are "fellow mortals," and that humans may choose to be vegetarians (herbivores), but we evolved to eat meat as well (omnivores). Why would Hudson want the reader to feel a kinship with the pig if it was just going to be eaten? How did you feel reading this part?

CLARENCE KING (1842–1901)

From MOUNTAINEERING IN THE SIERRA NEVADA (pp. 276–81)

1. Darwin had a huge impact on the study of geology, and we see the affects in how King discusses deep time, using terms like Paleozoic, Azoic, Triassic, and so on. These distinctions opened geology up as a field of study, much the same way Linneaus's system catalyzed the study of plant and animal life. How is Darwin's influence evident in King's work?

2. King and Muir were contemporaries. They both explored the rugged Sierra Nevada and wrote about its geological formation—especially theories of glaciation. Yet King attends more to explaining the scientific origins than does Muir (in the works presented here). Does King communicate concepts clearly? How do his purpose and style differ from Muir's?

3. As a mountaineer, King saw views that few people at that time had seen, and the larger perspective they gave him certainly helped in his studies of geology and geography. Think about times you've been at higher elevations, even in an airplane. How did your point of view from this height reveal patterns in the land? How did King's mountaineering help him understand and theorize about the forces that shaped the Sierra Nevada?

GERARD MANLEY HOPKINS (1844–1889)

From NOTEBOOKS AND PAPERS OF
GERARD MANLEY HOPKINS (pp. 281–86)

1. The editors refer to Hopkins's notion of "inscape" as his "concept for the integrity of every natural object or landscape." Using this definition, interpret what Hopkins means in one of his many passages mentioning "inscape."

2. Compare Hopkins's September 24 account of the northern lights to David Thompson's account of the aurora borealis (they are the same phenomenon). Thompson is fascinated by the relation of reason to sense, whereas Hopkins is interested in the relation of faith to sense. Compare their reactions.

3. According to Hopkins, "What you look hard at seems to look hard at you." Compare this assertion to Walt Whitman's claim to have "met and fused" with sea and shore. The two poets had much in common, but Hopkins lived in a more cynical era, when people were focused more on the practical. Many had grown disenchanted with Romantic views and even found nature to be an antagonistic force. To some degree, Darwin's theory of evolution weakened religious faith. People began looking to nature for science, not God. And yet Hopkins resists science's negation of faith. Where in his journal do you see evidence of this resistance?

RICHARD JEFFERIES (1848–1887)

From OUT OF DOORS IN FEBRUARY *and* ABSENCE
OF DESIGN IN NATURE (pp. 287–99)

1. Jefferies shows us how to reinterpret winter into a hopeful allegory. How does his writing reveal the "mystery of growth"? What must one do to learn of these mysteries?

2. Jefferies strongly believes that "a fixed observer usually sees more than one who rambles a great deal and covers ten times the space." Compare what Jefferies sees staying put to what King sees exploring the Sierra Nevada. Is Jefferies's close-up view better than King's panoramic perspective? Or does each offer vital information?

3. Jefferies states that only humans despair and only humans impose limits and order. The urge to classify living things often springs from a mechanistic view of nature which Jefferies abhors. He prefers the uncertainty of a "divine chaos." What does Jefferies gain and lose through this view?

MABEL OSGOOD WRIGHT (1859–1934)

From THE STORY OF A GARDEN (pp. 299–304)

1. Wright praises gardens in which the wild and cultivated land and species mix and merge, neither oppressed. According to Wright, what kind of give and take is necessary for this to occur?

2. Wright disdains gardeners who are too controlling and have "contempt [for] all wild things." But unlike Jefferies, she does support a blend of control and wildness. "Nothing was troubled but to free it from the oppression of some other thing," she says. "The gate is never closed except to violence." How would Jefferies respond?

3. There is a fairy-tale quality to some of Wright's stories. For instance, the garden was loved so much it gained a soul (Pinocchio?), and the woman goes about her day communing with the flowers and birds (Snow White?). Yet at other times Wright's voice is practical and full of advice. How does the blending of these different tones mirror and reinforce her message?

ERNEST THOMPSON SETON (1860–1946)

From WILD ANIMALS I HAVE KNOWN (pp. 305–12)

1. After reading "Lobo," how did you feel about Seton, given that the story was based on his experience as a bounty hunter? Seton did not make his character sympathetic to most readers. Why would he present himself this way?

2. Thoreau wrote in his journal, "If you have undertaken to write the biography of an animal, you will have to present to us the living creature, *i.e.* a result which no man can understand, but only in his degree report the impression made on him." Seton aligns himself

with Thoreau's view, opposing the mechanistic view of animals. How does Seton defend himself against the accusation that he is a "nature faker" who portrays animals as impossibly humanlike individuals? Use examples from Thoreau's journal to show why and how he would have supported Seton.

3. In defense of his animal stories, Seton carefully defines what is true in his stories and what liberties he has taken with the truth. (This is a familiar debate to anyone studying contemporary creative nonfiction.) Jefferies finds that what science supplies as truth is composed of "conventional agreements" made between members of a culture. "If you wish to really know the truth," Jefferies says, "there is none." How might Seton respond? How do you respond to their ideas of truth?

GENE STRATTON PORTER (1863–1924)

From THE MOTHS OF THE LIMBERLOST (pp. 312–20)

1. Stratton Porter is careful to avoid coming across as a Romantic. For instance, she says she "almost" thought God had felt sorry for her, and she does "not attempt to portray her feelings." Yet her approach is not mechanistic. How does she demonstrate her feelings for the creatures she studies?

2. Compare Stratton Porter's method of collecting and studying specimens with the methods of earlier naturalists who killed and dissected theirs. What role did photography play in her method and ethic?

3. Stratton Porter roamed outdoors and worked hard. She does not come across as timid or frail, yet one farmer suggests she lacks common sense and needs a guardian, and her husband commands their daughter to shade her mother rather than the moths. Though both these men likely acted out of kindness and concern, do you think they would have responded the same way to a man in those situations?

MARY AUSTIN (1868–1934)

THE LAND OF LITTLE RAIN (pp. 320–26)

1. "Not the law, but the land sets the limit," declares Austin. What examples does she give of the way humans must shape themselves to the desert? How is Austin's view different from Thaxter's and Wright's with their gardens?

2. Look at how King writes about the arid land beyond the Sierra Nevada, and compare his description to Austin's. How does their language reflect their different feelings about the desert?

3. Austin appreciates a landscape that few have championed. Hudson wrote of the Patagonia desert, which drew him in a way that kinder, inhabited landscapes did not. What makes Austin prefer the desert over other places? How does she try and make the reader see what she sees, feel what she feels?

LUTHER STANDING BEAR (1868–1939)

NATURE (pp. 326–31)

1. Standing Bear writes, "[F]or the Lakota there was no wilderness." *Wilderness* is a term western civilization uses to refer to nature outside of human control, the opposite of domestic space or home. Why did the Lakota have no use for this word? Why did the white man need it?

2. In Lakota culture, "the animal had rights," Standing Bear tells us. It is ironic that he should use the term "rights," the cornerstone of the white man's democracy, to explain Lakota views. How is this a strong rhetorical move?

3. Think of a nature writer who you think Standing Bear would agree had the "Indian point of view." Support your choice by comparing that writer's work to Standing Bear's description of Lakota beliefs and lifestyle.

EDWARD THOMAS (1878–1917)

Hampshire *and* The End of Summer (pp. 331–39)

1. Thomas describes a day that brings people happy, hopeful dreams. Yet we sense melancholy throughout the piece. How does he convey the melancholy?

2. Standing Bear says that the Lakota is "free from the fears of civilization." How are Thomas's "fears about immortality" related to the fears Standing Bear mentions? What does Thomas say causes those fears?

3. Thomas suggests some of the preoccupations of the Modernist movement, which critics cite as beginning with World War I. In these passages, there is a sense of alienation, impermanence, and existentialism. Existentialists believe that life and the universe are meaningless, Godless, and chaotic. Yet Thomas clearly finds some comfort in nature. How can nature, even if not divinely created, comfort Thomas in his "human desire of permanence"?

ROCKWELL KENT (1882–1971)

From N by E (pp. 340–42)

1. Kent's wilderness is a place where "peace predominates." Early views of wilderness were of a place where danger and fear predominated. How does Kent turn the tables on the connotations of wilderness and civilization?

2. Kent provides a nice contrast in tone to Powell, who also acknowledged the unknown but was more cautious than jubilant. Compare their accounts. If you were joining an expedition, which of these men would you want with you? Why?

3. Kent's account of Christian missionaries coming to Greenland is interesting because in one way he is the typical imperialist, glad that the "Gospel law" was brought to the "heathen folk." In another way, Kent is more open-minded, valuing the natives' ways as well and applauding their influence on the Christians. What are your reactions to this cultural exchange and Kent's response?

VIRGINIA WOOLF (1882–1941)

THE DEATH OF THE MOTH (pp. 343–45)

1. Woolf's tone is perhaps the most distant and formal of any nature writer so far. How does she achieve that distance? What affect does her tone have on you?

2. Like Thomas, Woolf (a Modernist) felt nature to be indifferent and without plan. Instead of focusing on other humans, as Thomas does, Woolf pities the day moth. Yet she also admires it. Thomas wrote, "This is the beginning of the pageant of autumn, of that gradual pompous dying which has no parallel in human life, yet draws us to it with sure bonds." Do Woolf's observations and response support or counter Thomas's statement?

3. Woolf's thoughts about the moth seem to reflect her view of life in general. Nature is "impersonal, not attending to anything in particular" (far from the comforting mother image). We are stuck where chance dropped us. How does this perspective strike you compared with others you have read? Is it preferable to anthropocentricism or Romanticism? Is it depressing or realistic? How does it affect your response to Woolf and her essay to know that she committed suicide by filling her pockets with rocks and walking into a river?

ISAK DINESEN (1885–1962)

From OUT OF AFRICA (pp. 346–48)

1. Dinesen suggests great contrasts in land and lifestyles between Africa and Europe. What differences is she emphasizing? Why?

2. Compare Dinesen's description of the air to Austin's. How does the focus on air reinforce the sense of rightness that each seeks to convey about their chosen places?

3. Dinesen speaks of the animals she's seen "in their own country" and finds reassurance in the fact that they are "out there still" at the time of writing. Yet throughout the piece there is a sense that the paradise being described is vanishing, like the American frontier that the Superintendent of the Census declared closed in 1890. Fresh New Worlds were being sought and explorers turned to Africa. Are new

frontiers being explored today? Are they the same kinds of frontiers as before?

D. H. LAWRENCE (1885–1930)

From FLOWERY TUSCANY (pp. 348–61)

1. In section I, what practices does Lawrence claim devastate and disfigure the land?

2. Compare Lawrence's description of the Tuscany farms to Wright's garden and its mix of the wild and the cultivated. Instead of a garden of a few acres, Lawrence discusses a whole countryside. How are the stakes changed?

3. Lawrence refers to the northern Europeans as people who believe "the phenomenal world is essentially tragical." Does this view remind you of Woolf, Thomas, Peattie, or other Modernist nature writers? In sections II and III, Lawrence, also a Modernist, expresses a different view: "The universe contains no tragedy, and man is only tragical because he is afraid of death." How does Lawrence's closeness to the flowers and land help him not fear death? How does his different understanding of what Jefferies calls "the mystery of growth" differ from that of the nature writers mentioned above?

L. GRANT WATSON (1885–1970)

UNKNOWN EROS *and* WAVE AND CLIFF (pp. 362–66)

1. Watson takes pleasure in watching the slugs copulate and in other "moments of ecstasy" in nature. Though he relates all he observes to "the realm of the invisible spiritual," Watson seems most interested in sensual, physical aspects of creatures' interactions in and with the world. How does this come across in his use of language and imagery?

2. "May it not be that these humble manifestations, composed [of nature], are the physical counterpart of our spiritual comprehensions?" Watson here calls to mind the Transcendentalists' "theory of correspondence" concerning nature and spirit, the interior and exte-

rior worlds. Compare Watson's rhetoric and views to those of Emerson and Thoreau. What does each look to nature for?

3. Watson praises slug love, a creature and act few might be inclined to watch, let alone rhapsodize over; Austin praises the desert southwest, a landscape few speak of longingly. Is it easier or harder to describe something few others appreciate? Do nature writers have an obligation to find and share the beauty and worth (even if spiritual rather than utilitarian) of unloved creatures and places? Why or why not?

HENRY BESTON (1888–1968)

From THE OUTERMOST HOUSE (pp. 366–75)

1. Beston gives many images of unity, of each creature working instinctively and naturally in concert with others. Though Beston refers to "the collective self" of flocks, he emphasizes that animals are not machines or tragic, but are individuals complete in themselves. In a time when many speak of feeling alienated from other humans, and nature, why is this emphasis so important?

2. *Ritual* is a favorite term of Beston's in this piece. As far as seasons and lives, "they come, they go, they melt away, they gather again." Compare his view of the natural cycles with D. H. Lawrence's, especially the last paragraph of Section III of "Flowery Tuscany."

3. Beston explains that at night "we have a glimpse of ourselves and our world islanded in its stream of stars." When darkness reveals the larger universe and all we do not know, we may feel uneasy at the possibility of alien life forms with consciousness equal to or greater than ours. The change of scale forces us to consider that not all the universe was made for humans to control. Is space our final frontier, as *Star Trek* suggested? What happens to our perspective of nature when we consider the larger universe? Are other planets a part of nature? Will there someday be field guides to Mars?

ALDO LEOPOLD (1888–1948)

From A Sand County Almanac (pp. 376–97)

1. Leopold's history of the marsh does not imply that working the land is bad, but that certain ways of working the land are harmful. What attitudes and actions does he condemn? Why?

2. In "Thinking Like a Mountain," Leopold reveals how he used to shoot wolves when there was a bounty on them. Compare this essay with Seton's "Lobo." What realizations did both men have? What does Leopold recognize about the relationship between wolves and the land?

3. How *does* one think like a mountain, according to Leopold? Is it a matter of not judging nature by "whatever human values are in style," as Beston says? Discuss a couple of passages from "The Land Ethic" that illustrate how difficult it is for people in western civilization to think like a mountain. Does Leopold give you hope that we can learn? Why or why not?

JOSEPH WOOD KRUTCH (1893–1970)

Love in the Desert (pp. 397–410)

1. Krutch wryly applies human aesthetic and spiritual standards to plant and creature reproduction. How does Krutch use humor as a rhetorical tool to convey his message?

2. Many Romantics looked to nature for moral instruction. Krutch severely critiques that approach. What are his strongest arguments? How would Krutch react to Watson's interpretation of slug sex?

3. Despite his distaste for extreme Romantic notions, Krutch defends the idea that animals have emotions and bases his reasoning upon evolutionary science. He juxtaposes his view with that of the average cynical civilized person of his time, who avoids at all costs the pathetic fallacy—that is, attributing human emotions to external objects (including animals), supposedly indicative of a sentimental rather than realistic view of the world. But cultures such as the Lakota that Standing Bear describes believe that all things have spirit. How does Krutch negotiate a middle ground? In what way does he believe that humans differ from other animals?

HENRY WILLIAMSON (1895–1977)

From TARKA THE OTTER (pp. 410–15)

1. This is the story of an animal's life in a habitat encroached upon by humans. What does Tarka face that his ancestors did not? How does it affect him?

2. In a long narrative, Williamson conveys the mating habits of the river otter. Compare his description of Tarka and White-Tip, their feelings and intents, with Krutch's views of animal behavior. Would Krutch approve of Williamson's account? Why or why not?

3. The drama of animals mating is equal to the drama of the predator/prey relationship. It is hard not to feel somewhat uneasy if a male of some species is forcing himself on an unwilling partner. Especially in this age of gender awareness, studying animal behavior for guides to human conduct can be stimulating, illuminating, horrifying, and dangerous. Many females of other species mate "against their will," so to speak, and many of them kill and sometimes eat the male after (or during) copulation. How has reading Williamson and Krutch (and others) broadened, challenged, or confirmed your perspective on male and female interactions?

DONALD CULROSS PEATTIE (1898–1964)

From AN ALMANAC FOR MODERNS (pp. 415–22)

1. Many of Peattie's observations on human significance are based upon his understanding of the biotic pyramid that Leopold refers to in "The Land Ethic." The base of the biotic pyramid comprises soil and microorganisms. The largest, most complex creatures are at the top— fewer in number, like a pyramid. What does Peattie mean when he speaks of bacteria as the lords of creation, "whose subtraction would topple the rest of life"? And what does he mean when he asks, "[O]f what use, pray, is man? Would anybody, besides his dog, miss him if he were gone?"

2. Peattie believes that only one conclusion can be drawn from the realistic observation and investigation of nature. He speaks condescendingly of people who choose to insert soul, purpose, and symmetry into the cosmos because to do so is comforting. Compare Peattie's

view (outlined at the end of *March* 31) to Krutch's view (outlined in his final paragraph).

3. Peattie presents the evolution of our understanding of nature, from the Egyptians and Greeks through Aristotle to the present. "[H]ow can a man base his way of thought on Nature and wear so happy a face?" he asks. Is he sincere? Is he sincere in trying to find an answer while being a "courageous thinker" and accepting "the stern fact that the universe is hostile"? Or does Peattie have no hope of finding happiness in a realistic "philosophy based upon nature"? How does his language reveal his beliefs?

VLADIMIR NABOKOV (1899–1977)

BUTTERFLIES (pp. 423–32)

1. Why does Nabokov think the "mysteries of mimicry" go beyond what Darwin's theories can explain?

2. Compare Nabokov's stories of specimen hunting and collecting to Stratton Porter's, whose lifelong hunt for a Cecropia was spurred by her dismay over letting one escape when she was a girl. How are their personalities and obsessions different? What role does memory play for each of them?

3. Nabokov declares, "I discovered in nature the nonutilitarian delights that I sought in art. Both were a form of magic, both were a game of intricate enchantment and deception." For both the novelist and the butterfly collector, the act of naming is artistic, magical, and informative. How is the urge to discover (as in butterfly collecting) similar or different from the urge to create (as in writing fiction)? How do these urges shape human relations to the natural world?

SIGURD OLSON (1899–1982)

NORTHERN LIGHTS (pp. 432–35)

1. In this skeptical modern age, how does Olson make his argument that experience (not superstition or Romanticism) can contradict or refute science?

2. Olson recalls his childhood experience of the northern lights. Compare his discussion of the change in perspective that has come with maturity and knowledge to Clemens's memory of his first sunset over the Mississippi River. Do they have the same tone? Would they agree that such wonder can never be recaptured?

3. Olson's discussion of his thoughts that night suggests that science can convey reasons or causes behind phenomena, but only art can convey the experience of such phenomena. Is this why Beston said, "[P]oetry is as necessary to comprehension as science"? Do you think this is true? Respond to Olson's question of whether the "legendry [can] survive scientific truth."

EDWIN WAY TEALE (1899–1980)

From THE LOST WOODS (pp. 435–39)

1. Teale says that the Lost Woods were a symbol. What did they symbolize? Why did that patch of woods have such significance for him?

2. Sigurd Olson says there is "wonderment only a child can know." Do you think Teale's wonder over the Lost Woods would diminish if he saw it a second time?

3. Childhood memories play a great role in the works of twentieth-century writers such as Nabokov and Olson. Williamson wrote stories to introduce children to the natural world, demonstrating a desire to instill a reverence for nature. How might historical context explain this trend? Luther Standing Bear said the main difference between Lakota culture and western civilization is the way children are raised. Do you agree? What does the work of the writers mentioned above directly or indirectly reveal about how children are raised in our culture?

E. B. WHITE (1899–1985)

A SLIGHT SOUND AT EVENING (pp. 440–48)

1. What does White mean in the first paragraph by "that language we are in danger of forgetting"?

2. White praises *Walden* as "the best youth's companion yet written by an American." Why is that? Would Luther Standing Bear agree or disagree?

3. White wrote this piece not long after the first atomic bombs were dropped "in the brooding atmosphere of war and the gathering radioactive storm." One hundred years after Thoreau published *Walden*, the world has changed drastically. Why does White still find *Walden* "pertinent and timely"? What can nature writing accomplish in a time when "ten thousand engineers are busy making sure that the world shall be convenient even if it is destroyed in the process, and others are determined to increase its usefulness even though its beauty is lost somewhere along the way"? Use examples from your favorite nature writers to explain your response.

MERIDEL LeSUEUR (1900–1996)

THE ANCIENT PEOPLE AND THE NEWLY COME (pp. 448–52)

1. LeSueur's striking imagery manages to combine human history with natural history—two timelines usually kept separate. How does she do this, and how is it radical?

2. "One was born into space," says LeSueur of children born on the Great Plains at the turn of the century. One is also born into a body, and throughout LeSueur's writing, images of human bodies and the land intertwine. How does she do this and why?

3. While LeSueur believes that humans are part of the natural world, she is also aware that most humans—especially the pioneers—see themselves as separate, with nature as a frequent enemy. Preparing for winter on the plains was "like preparing for a battle." This push-pull tension is conveyed through imagery and stories that depict how people "flowed through and into the land" while also being "evicted, drouthed out, pushed west." Which force is stronger? If our bodies are part of the land, as LeSueur maintains, then what are the consequences of battling the land?

RENÉ DUBOS (1901–1982)

A FAMILY OF LANDSCAPES (pp. 453–57)

1. The term *natural* can have many different meanings. What does Dubos mean by *natural*? How can you tell?

2. "Ecology becomes a more complex but far more interesting science when human aspirations are regarded as an integral part of the landscape." With this statement, Dubos takes Thoreau's and others' notions about humans as "a part and parcel of nature" to another level. How would Thoreau react to Dubos's essay? Are "human aspirations" utilitarian or spiritual or both?

3. Dubos's discussion of the role the altered Greek landscape has played in the development of art, poetry, and philosophy complicates our understanding of aesthetics, which, up to now, has almost exclusively referred to what is natural or what mimics the natural. When, if ever, can the unnatural be beautiful? When, if ever, are beauty and inspiration reasonable justifications for altering a landscape? Explore your views on this subject by reacting to Dubos's essay and referring to other nature writers who deal with aesthetics and land management.

NORMAN MACLEAN (1902–1990)

From A RIVER RUNS THROUGH IT (pp. 457–65)

1. Maclean plays with perspectives: the view each brother has of the other, the view each brother has toward other creatures, and the views the jackrabbit and trout have of the world and the brothers. In what ways does Maclean show his self-consciousness about trying to imagine what another is thinking? What effect does this awareness bring to the piece?

2. How is Paul's sense of companionship with the jackrabbit different from and similar to Muir's relationship with the water-ouzel? Consider the time period, the sensibilities of the two men, and the different circumstances surrounding the interactions.

3. Maclean's narrative of the brothers fly fishing conveys the interconnection of the brothers, the fish, the river, and the rocks. He also describes the how-to of fly fishing in detail, similar to the way

Clemens explained how a steamboat captain reads a river. Still, the beauty is luminous and the experience all-absorbing. How does Maclean merge utilitarian and aesthetic descriptions? Does the fact that this piece, though nonfiction, is written as a story and not as an essay affect your experience as a reader?

JOHN STEINBECK (1902–1968)

From THE LOG FROM THE SEA OF CORTEZ (pp. 465–68)

1. Steinbeck asserts that "books of fact" are crafted and controlled as much as books of imaginative literature. What does he mean by "all knowledge patterns are wrapped, first, by the collective pressure and stream of our time and race, second by the thrust of our individual personalities"?

2. There is no pure scientific impulse divorced from other personal desires, Steinbeck suggests. Nabokov acknowledged that there was more to his pursuit of butterflies than just service to science. How would Nabokov answer Steinbeck's question:

   Why do men, sitting at the microscope, examine the calcareous plates of a sea-cucumber, and, finding a new arrangement and number, feel an exaltation and give the new species a name, and write about it possessively?

3. Steinbeck openly challenges the idea that the human observer can be objective—that is, have no interaction with or influence (either emotional or physical) on what is being observed and how it is being recorded. In this age of skepticism and reliance upon practical, verifiable truth, this view is threatening and highly controversial. Yet Steinbeck does not want to abandon "objective" observations and knowledge. Do you think it is possible or worthwhile to combine objective and subjective approaches?

LAURENS VAN DER POST (1906–1996)

From THE HEART OF THE HUNTER (pp. 468–74)

1. Ben comments that "the gembuck is royal game here and protected, and the Bushman is not." What does this statement mean, and why did Van Der Post include it?

2. Van Der Post found Africa to be "a holy world of quest." Compare his reflections near the end of his trip with Steinbeck's musings before heading off to the Sea of Cortez. Would they have been compatible travelers?

3. Consider Ben's description of all the ways African birds impress him. What is his tone? Reverent? Sentimental? Overblown? Silly? How do you react to his comparisons between different animals? Why?

T. H. WHITE (1906–1964)

THE SNAKES ARE ABOUT (pp. 474–79)

1. T. H. White consciously plays language and imagery off the familiar Christian image of evil snakes, such as Satan tempting Eve in Eden. How and why does he do this?

2. Compare T. H. White's description of communing with snakes to other writers' descriptions of their relationships to individual animals. Think of Maclean and the trout, his brother and the jackrabbit, Muir and the water-ouzel, and so on. On the spectrum of pets to wild animals, where does each fall? What are the different characteristics of each bond?

3. Why does T. H. White feel that accustoming the snakes to human handling will "steal them from themselves"? What is the line between communing with a wild creature and domesticating it? Are there times when T. H. White crosses or straddles this line? Why is he so concerned about it?

RACHEL CARSON (1907–1964)

THE MARGINAL WORLD (pp. 479–85)

1. How is the *hydroid tubularia* like the shoreline, with its double life above and below the saltwater? Where else do images and themes of doubleness surface in Carson's writing?

2. Carson explains how evolution has shaped different shorelines, how "the past and the continuous flow of time" reveal themselves. Compare her account of the changes she reads in the mangrove island with King's geologic description of how the Sierra Nevadas were formed. How does Carson's inclusion of creatures affect her account?

3. "Creation is here and now," Beston said, and Carson agrees, finding evidence of more than destruction in the modern landscape. Yet she does not gloss over destruction. Carson's wish to see the flamingos Audubon witnessed reminds her of the species that are no longer there. Have you been to any places where you felt the larger forces of evolutionary time? How did those places evoke this feeling? How does Carson's view challenge the commonly held idea that the human species is the climax of creation?

LOREN EISELEY (1907–1977)

THE JUDGMENT OF THE BIRDS (pp. 485–93)

1. Eiseley elaborately strings together varied encounters in nature, layering the experiences in order to satisfy "the human need for symbols." What do birds symbolize in this essay and in our culture?

2. Even in New York City "there are patches of wilderness," Eiseley says. Compare his experience with the pigeons to Orwell's description of the pleasures of spring in the London slums. What do these men achieve by focusing on urban nature? How does looking at nature in an urban context affect your understanding of wilderness?

3. Eiseley presents these encounters so that evolutionary and personal perspectives are intertwined. He admits to being a brooding loner and seems to look for hopeful signs in the wilderness. His moments of revelation afford him a larger perspective that he would like to pass on "to those who will fight our final freezing battle with the void."

How do you react to Eiseley's certainty that there will be an end to the human species? Is it better for nature writers to record, not define, marvels? Does Eiseley follow his own advice?

RICHARD WRIGHT (1908–1960)

From BLACK BOY (pp. 494–96)

1. Wright uses a list to express his early impressions of the world. What does he include in his list, and what does he exclude?

2. Using single-sentence descriptions of childhood memories, Wright makes no distinction between natural phenomena and his responses to them. Compare Wright's weaving together of the human and natural worlds with LeSueur's intertwining of human and natural history. How does each writer's description provide information about his or her childhood?

3. Wright gives us only one image that reveals his boyhood experience of prejudice. What is the effect of including it in this list? What does race have to do with one's experience of nature? How might it affect both the writer and the reader of nature writing?

ARCHIE CARR (1909–1987)

THE LIVELY PETES OF PARQUE VARGAS (pp. 497–504)

1. In addition to the love life of sloths, Carr is interested in what scientists call the carrying capacity of the park—in this case, how many sloths the laurel trees can support. Why is Carr so fascinated by the balance the two species have struck?

2. Carr calls the population balance in the park a "natural equilibrium." What does he mean by "natural" in this context? How is it similar to or different from what Dubos means by "natural" in his essay?

3. The modern understanding of evolution informs Carr's musings on the sloths' strange habits. Instead of shaking his head at their silly behavior, he assumes their slowness is a "useful adaptation." Following Carr's line of thought, what reasons can you think of for the sloths' evolving in such a way? What might be the evolutionary

advantage to humans' tendency to wonder and pursue knowledge? What might be the disadvantages?

WALLACE STEGNER (1909–1993)

From GLEN CANYON SUBMERSUS and
CODA: WILDERNESS LETTER (pp. 504–19)

1. Stegner argues for the protection of certain areas to guard their spiritual resources. In what ways are silence and the wilderness idea spiritual resources? What else does Stegner imply we will lose if we lose them?

2. Stegner describes a landscape altered by "human aspirations," to use Dubos's term. Many people now vacation on Lake Powell, as people once enjoyed Glen Canyon. How is this situation similar to and different from the one Dubos describes in Greece? Whose aspirations should be fulfilled—those of the people who prefer Lake Powell or who prefer Glen Canyon?

3. Expanding the idea of what "kinds of wilderness [are] worth preserving," Stegner adds prairies and deserts to mountains and forests. Still, there is a sense here, however unspoken, that only western landscapes fit the definition of wilderness. Even Thoreau privileged the west over the east when he went walking, because the open spaces symbolized freedom and the future. What is your idea of wilderness? What types of land should be protected from human encroachment? Would you add other landscapes to Stegner's list? Why or why not?

JACQUETTA HAWKES (1910–1996)

From A LAND (pp. 520–25)

1. "In the history of thought, this is the age of history," states Hawkes, and goes on to use geologic imagery to describe the forming of human consciousness. Why does she do this? What "other forms of memory" is she pursuing?

2. Hawkes praises knowledge and curiosity but laments the way that we have "allowed [facts] to become our masters." She chooses metaphor

over facts to relate change and form. How does her approach to her book agree with Steinbeck's introduction?

3. How does Hawkes's belief that creation or evolution is inevitable and predetermined allow her to experience awe more easily than terror, humility more easily than insignificance? What similarities and differences exist between her perspective as an archeologist and Eiseley's as an anthropologist? How does each talk about memory and the future?

JOSEPHINE JOHNSON (1910–1990)

From THE INLAND ISLAND (pp. 525–32)

1. In what way does Johnson's alternation of choppy rhythms with smooth sections affect the mood of the piece? What does she achieve with this technique?

2. Compare Johnson's essay with Orwell's. How does each begin? At what point and in what way do the political concerns get revealed? "How can I live in two worlds?" asks Johnson. "Is it wicked to take pleasure in spring, and other seasonal changes?" asks Orwell. How would each respond to the other's question?

3. The headnote states that Johnson's awareness of the Vietnam War "caused her to question the validity of her own solitude and peace." A similar tension surrounds the wilderness movement today. Why should we protect wilderness for the upper and middle classes to vacation in while those less economically fortunate—often minorities—are stuck in the polluted inner city or depressed rural towns? Is it enough, as Stegner said, for people who cannot get to wilderness to contemplate the idea of it?

LEWIS THOMAS (1913–1993)

DEATH IN THE OPEN *and*
THE WORLD'S BIGGEST MEMBRANE (pp. 533–38)

1. Why does Thomas say, "[W]e will have to give up the notion that death is catastrophe, or detestable, or avoidable, or even strange"?

2. Compare Thomas's explanations of evolutionary concepts with Car-

son's and Hawkes's. Which do you find clearer or more thought-provoking? Which makes better use of metaphor? Which is more hopeful? Use examples from the texts to support your opinions.

3. Thomas points out that our culture does not engage with the dying. Human death, like human waste, is a part of natural process that seems distasteful—maybe because it reminds us that we are animals. Still, Thomas admits, if all the dying were out in the open it would be overwhelming. "But [hiding it] does make the process of dying seem more exceptional than it really is," he continues, "and harder to engage in at the times when we must." What do you think of our culture's attitude toward death? Is there a connection between how we treat human death and our destruction of creatures and habitats?

JOHN HAY (b. 1915)

The Common Night *and*
The Dovkie and the Ocean Sunfish (pp. 539–45)

1. Through observations of the alewives and dovekies, how does Hay link Cape Cod to the global world? How is his local knowledge like the Eskimo's?

2. "How can we predators dare count the bodies, human and other than human, of those on whom we prey!" chastises Hay. Read the paragraph that begins with this statement and relate it to Leopold's "Thinking Like a Mountain" and "The Land Ethic." Hay distinguishes between daring to do something and making money by doing something. What does Leopold say about the role of economics and finances in our treatment of the environment? How does Leopold answer Hay's last question in that paragraph?

3. Hay speaks of "the cruel mother that might take care of [the dovekie] better than I." LeSueur and others have used the metaphor of the earth as mother. But the earth they describe is not only protectively maternal but also "cruel," practicing what psychologists call "tough love." If the earth is a mother and not indifferent, how do we explain the pain caused by natural forces? What do we risk when we give in to our urge to care for a wounded creature? How do we assess the consequences? Is the urge to interfere an animal instinct toward caring for other life, or is it an example of our arrogance and need for control?

THOMAS MERTON (1915–1968)

RAIN AND THE RHINOCEROS (pp. 545–53)

1. According to Merton, how do "they" (city dwellers) decide what has value?

2. Lewis Thomas critiques our culture's lack of engagement with death. Merton says that in order to be awake, to have identity, a person must "accept vulnerability and death." Are their goals for getting others to face death similar? Why or why not?

3. Merton states that the city is the realm of mystification, created and continued via "its own myth," nothing but a "fabricated dream" and "a world of mechanical fictions." One of the myths of the city is that in it all our needs can be immediately fulfilled. What does Merton suggest are the results of believing in such a myth?

FAITH McNULTY (b. 1918)

MOUSE (pp. 554–60)

1. In describing Mouse, McNulty says, "[S]he had fine sharp teeth and a striking air of manful competence." The contrast between the pronoun and characteristic is a bit jolting. Where else does McNulty's narrative evoke an awareness of gender and gender stereotypes?

2. Compare the McNultys' response to Mouse with Hay's encounter with the wounded dovekie. Was it any better for the McNultys to interfere than for Hay? Then compare their subsequent relationship to Mouse with T. H. White's relationship to the snakes. Once we handle a wild creature, have we upset its wild life so much that we are obligated to care for it? Should McNulty have let Mouse go? Should T. H. White have kept and protected Matilda?

3. McNulty studied Mouse, even going so far as to read a veterinary manual to try to understand the strange lump, but Mouse was clearly a pet and not a naturalist's specimen. What distinguishes the treatment of a pet from the treatment of a "specimen"? What motivations guide a naturalist and a pet owner in relation to animals? Why was the doctor who suggested an autopsy so out of line?

FARLEY MOWAT (b. 1921)

From NEVER CRY WOLF (pp. 561–66)

1. Mowat goes to great lengths to describe how foolish he must have appeared, how undignified. Why is it important to his story that he make us see him this way?

2. Eiseley said that it is better for nature writers to "record their marvel, not to define its meaning." The "human concept of wolf character" that Mowat inherited came from previous generations when the meaning of the wolf was defined in stories. Are Mowat and Eiseley wary of definitions for the same reasons?

3. Remember how Clare ranted against those who "trust to books and repeat . . . the old error"? Tales like "Little Red Riding Hood" and "Peter and the Wolf" could be considered such books, although the fear of wolves that the stories encourage certainly was instructive at one time. Most wolf attacks on humans were in times and places of war, epidemics, and famine. Perhaps the abundance of unburied corpses caused some wolves to overcome their natural fear of humans. But as circumstances have changed, so, implies Mowat, should our stories. Rewrite one of the tales cited above to convey a new concept of wolf character.

JOHN HAINES (b. 1924)

MOMENTS AND JOURNEYS (pp. 566–72)

1. Like Eiseley, Haines links separate scenes that are etched in his memory. How does his narrative reflect both "the dream journey and the actual life"?

2. Haines, at least while at his homestead in Alaska, lived the kind of life Merton recommends. What were Haines's needs? In what ways are the two writers' philosophies about solitude, cities, and other things alike?

3. Like Merton, Haines describes an awareness of both interior and external realities—an awareness they see disappearing from modern, urbanized humans. Yet Eiseley and Orwell demonstrate this awareness in New York City and London. The recognition of seasonal rit-

uals can occur in urban places as well, though it may be harder as Merton and Haines claim. Think about your home, your school, or your city. What seasonal rituals do you associate with the place? For instance, a student from Florida marks the arrival of winter as when the U.P.S. man starts wearing long pants.

MAXINE KUMIN (b. 1925)

SILVER SNAFFLES (pp. 572–76)

1. Many nature writers have referred to the verbal communication or language of other creatures or aspects of nature—even water and wind. In discussing the horse–human bond, Kumin stresses another type of communication as well. What is it, and why is it significant?

2. Kumin refers to the "intrinsic worth" of caring for and working with horses. Compare the intrinsic value she describes with the instrumental value Merton says most city dwellers assign to things that can be sold or used to make money. What is the difference? Why is the intrinsic worth of animals and landscapes rarely recognized in our society?

3. Kumin admits that she risks sounding sexist when she outlines why women tend to work best with problem horses. Did you feel she was being sexist? If women are more empathic, subtle, nurturing, and instinctive, as she claims, is it the result of biological (essential) or cultural (learned) influences? Have you noticed differences between how men and women write about animals and nature? Use textual examples to support your opinion.

ANN HAYMOND ZWINGER (b. 1925)

OF RED-TAILED HAWKS AND
BLACK-TAILED GNATCATCHERS (pp. 577–86)

1. Like Austin's, Zwinger's desert narrative revolves around water. But unlike Austin's, many of Zwinger's details and images are domestic. Find some examples of such details and images, and analyze how they shape her story of spending time alone in the desert. Would you respond differently if her images had been more rugged and wild?

2. "I prefer the absences and big empties," says Zwinger, explaining why lush and/or closed landscapes do not appeal to her. Based on the other nature writers you have read, does there seem to be a preference for open, "seamless" landscapes over closed, "easy" ones? Which other writers would align themselves with Zwinger? Which would disagree? What type of landscape do you prefer?

3. Zwinger begins by wondering if she will be "uneasy alone." Many male nature writers have contemplated the importance and effect of solitude, as Zwinger's quotation from Krutch demonstrates. Though there have always been women who cherished solitude and ventured alone into the wild, their number in western cultures has consistently been much smaller than that of men. In fact, in America the valuing of wilderness arose in part because it was where boys could learn to be men and overly civilized men could renew their American character and virility. What might have been some of the reasons women were discouraged from venturing alone outdoors? Were any of these reasons valid? Are any still used to discourage women from taking Thoreauvian "excursions"?

J. A. BAKER (b. 1926)

BEGINNINGS (pp. 587–91)

1. "Be alone. . . . Learn to fear. To share fear is the greatest bond of all," instructs Baker. Why is the attitude expressed by this quotation so different from that of most early explorers and naturalists? Why is it that "the hunter must become the thing he hunts"?

2. Baker's prose is startlingly evocative and fresh like LeSueur's. Both mention fear and an intimate connection with their subject—the hawk and the Great Plains. But the wildness of their poetic language is perhaps their strongest similarity. Examine how each uses syntax, rhythm, metaphor, story, and forceful statements. What does such an examination help you discover about their methods and styles?

3. Baker describes his part of England as "profuse and glorious as Africa." But his tone is mournful as he notes the summer woods full of dying birds poisoned by "filthy, insidious . . . farm chemicals." DDT, the pesticide that Carson spoke out against in *Silent Spring*, almost wiped out the peregrines, but efforts since have helped the species rebound. Leopold said, "[W]e can be ethical only in relation

to something we can see, feel, understand, love, or otherwise have faith in." Can nature writing help make this emotional and ethical connection between people and creatures or habitats that they might never see? Why or why not?

JOHN FOWLES (b. 1926)

From THE TREE (pp. 592–605)

1. Fowles defines nature as "an experience whose deepest value lies in the fact that it cannot be directly described by any art . . . including that of words." Yet he says that "the key to my fiction . . . lies in my relationship with nature." How does he explain this?

2. Fowles laments the shift from a humanistic approach to a Linnaean or mechanistic view of nature. Examine Fowles's views through the lens of Steinbeck's comments on objectivity. How do Fowles's statements reflect Steinbeck's beliefs? In what ways does Fowles go further in his critique?

3. Fowles speaks of curing the world's ills, of nature as therapy, of the exodus of people from rural to urban areas. Fowles, Johnson, Merton, Haines, and Baker write partially in response to a world in crisis. How does this affect the tone, mood, and rhetorical choices of these writers? How does it affect your response as a reader?

FRANKLIN RUSSELL (b. 1926)

From THE ISLAND OF AUKS (pp. 605–13)

1. What emotions do Russell's descriptions evoke in you? Why do you think he wanted to return to the island and "live on it in order to turn [his] disbelief into lasting memory"?

2. How does Russell's writing illuminate the following statements?

> It speaks of the return of life, animal life, to the earth. It tells of all that is most unutterable in evolution—the terrible continuity and fluidity of protoplasm, the irrepressible forces of reproduction,—not mystical human love, but the cold batrachian jelly by which we vertebrates are linked to the things that creep and writhe and are blind yet breed and have being.   —DONALD CULROSS PEATTIE

We will have to give up the notion that death is catastrophe, or detestable, or avoidable, or even strange.   —LEWIS THOMAS

There is a kind of coldness, I would say a stillness, an empty space, at the heart of our forced co-existence with all other species of the planet. Richard Jefferies coined a word for it: the ultra-humanity of all that is not man . . . not with us or against us, but outside and beyond us, truly alien. It may sound paradoxical, but we shall not cease to be alienated . . . from nature until we grant it its unconscious alienation from us.   —JOHN FOWLES

3. What role do Russell's associations play in his first visit to Funk Island? Describe the memories he recalls. How does he use them to help sort through everything he hears, sees, and smells? What does Russell mean when he says, "only in retrospect could the island become real"?

EDWARD ABBEY (1927–1989)

THE SERPENTS OF PARADISE *and*
THE GREAT AMERICAN DESERT (pp. 614–27)

1. Abbey begins "The Serpents of Paradise" by posing questions about the human–animal relationship with irreverent humor. How do his humorous statements highlight the complexities behind the issues?

2. *Anthropomorphism* is the practice of ascribing human form or attributes to animals, plants, other nonhuman presences—a practice realists frown on as much as they do the pathetic fallacy. Which writers other than Abbey have practiced anthropomorphism? Which ones do so in a Romantic way, and which recognize that animals have "beautifully selfish reasons of their own"?

3. Abbey buddies up to the reader, using a casual tone and making fun of himself, to help balance his commanding directives. First, he tells us to stay out of the deserts. Then he urges us to join an army of hikers fighting to save "what wilderness is left in the American Southwest." In fact, in "The Great American Desert" he pretends his purpose is to inform us how to survive in the desert when his real goal is to tell us how to help the desert survive. What threats does the desert face? What advice does Abbey give those willing to fight to save it? How does he make the transition from warning us away to wooing us into loving land so "spare, rough, wild, undeveloped, and unbroken"?

PETER MATTHIESSEN (b. 1927)

From THE TREE WHERE MAN WAS BORN and THE WIND BIRDS (pp. 664–79)

1. Matthiessen often reflects on the population balance or carrying capacity of certain species and regions. What specific issues concern Matthiessen?

2. Matthiessen describes the lions as having "some dim intuition that the time of the great predators was running out." Why does he feel such empathy for these predators? Remember the "Land Pyramid" in which Leopold explains the place humans and lions occupy. How is the lions' situation similar to that of the African people?

3. Conservation is an imported ethic in many third world countries. Some claim it is a part of western imperialism and as problematic as imposing Christianity on other cultures. "The tourist industry," says Matthiessen, is "the last barrier between the animals and a hungry populace." How is this reliance on tourism for conservation problematic? Remember Leopold's land ethic. Do developed countries have a right to pressure poorer nations into protecting their natural resources when their citizens are going hungry? What are the alternatives?

NOEL PERRIN (b. 1927)

From PIG TALES (pp. 644–50)

1. Perrin, like Abbey, could be accused of anthropomorphism. Abbey defended himself directly. How does Perrin implicitly defend himself in these stories? What rhetorical techniques does he use to establish his credibility?

2. Perrin, like Crèvecoeur, is a farmer and acts as the law in the barnyard. Contrast Perrin's account of the little boy being punished for questions about the pigs with Crèvecoeur's account of taking his son to exterminate ants stealing honey from bees. How does each story reflect its time period?

3. What acted as the law in Mr. Harrington's pig barn? Why does Perrin prefer "life in a pig barn to life out here in the big wide world"?

Crèvecoeur says that "one species of evil is balanced by another." Does that apply to humans as well? Is having a sense of justice limited to humans? Why, or why not?

URSULA K. LE GUIN (b. 1929)

A VERY WARM MOUNTAIN (pp. 651–57)

1. Le Guin compares the president of the United States to Mount St. Helens. What point is she making about power and scale?

2. Instead of people trying to think like an erupting mountain, Le Guin says, "everyone takes [the eruption] personally." How does this essay make readers engage death and face the ways people try to process or deny events like volcano eruptions and earthquakes? What would Lewis Thomas say about the different reactions Le Guin recounts? In what way would he like us to respond?

3. Seeing the earth as mother and land as woman goes far back in time and across cultures. Le Guin puts a new spin on the metaphor, seeing Mount St. Helens as a sister and interpreting the eruption as an act of "unmistakably feminist solidarity." Le Guin's perspective is that of an ecofeminist. Ecofeminism, which emerged in the 1970s, is the belief that in our culture women, land, and animals have been oppressed in similar ways and for similar reasons. Ecofeminists believe that all oppressions are interconnected. How does Le Guin's essay show the relationships she finds between the treatment of and attitudes toward Mount St. Helens and women? What is your reaction to this perspective?

EDMUND O. WILSON (b. 1929)

THE BIRD OF PARADISE (pp. 658–62)

1. Wilson says that "because biology sweeps the full range of space and time, there will be more discoveries renewing the sense of wonder at each step of research." How can wonder increase with knowledge? How can research expand ignorance as well as understanding?

2. Steinbeck wanted to merge the two approaches of science and art,

objectivity and subjectivity. Where do he and Wilson agree? Where do they disagree?

3. Wilson cites the major complaints humanists have against science: "it reduces nature and is insensitive to art, that scientists are conquistadors who melt down the Inca gold." How does Wilson defend scientists? How does his imagery redefine the laboratory scientist as more than a cold and distant observer? What dynamic does he believe exists between the humanities and science? Do you think Wilson successfully blends these two approaches in this essay? Why or why not?

GARY SNYDER (b. 1930)

ANCIENT FORESTS OF THE FAR WEST (pp. 662–83)

1. Snyder combines storytelling, technical instruction, polemic, poetry, and philosophy into a dense essay on ancient forests. How and when does he switch from one mode to another? Which do you like the most? Which were harder to understand? What links them together?

2. Snyder is a bioregionalist, one who inhabits a place defined not by political boundaries but by watersheds and ecosystems. The bioregional community is the same as Leopold's land community. "The less violent the manmade changes, the greater probability of successful readjustment" in the ecosystem, claims Leopold. Why does Snyder feel better about selective cutting than clearcutting? What does he say is the difference between renewable and sustainable resources? How can our actions make a renewable resource no longer sustainable?

3. Snyder starts by describing a Western Red Cedar that he thought of as an advisor, then says he was "instructed by the ghosts of those ancient trees" in the same way he was instructed by his uncles. Gradually he builds the controlling metaphor that "the human community, when healthy, is like an ancient forest." How does he complicate the metaphor by incorporating his past as a logger? By expanding the scope from forests of the west to global forests? How are people turned into commodities like trees? What do these ancient forests, or elders, have to teach us?

JOHN McPHEE (b. 1931)

UNDER THE SNOW (pp. 684–90)

1. As Robert Finch and John Elder note, McPhee does not state his opinions overtly but communicates them through "careful choice of fact and detail." Why, then, does he include all the comparisions between eastern and western bears?

2. How does McPhee's tale of interfering with black bears compare with McNulty's story of Mouse or Hay's experience with the dovekie?

3. Wildlife management introduces a level of interference with wild animals. Do you think such management is necessary or justified? Are we correcting our mistakes or extending them? Does wildlife management make bears and other animals less wild?

EDWARD HOAGLAND (b. 1932)

From HAILING THE ELUSORY MOUNTAIN LION
and THOUGHTS ON RETURNING
TO THE CITY . . . (pp. 690–706)

1. Unlike most nature writers who compare the city to the country, Hoagland holds them in equal standing. What positive points does he make about New York City? How is it like the mountain where he lives the rest of the time?

2. Compare Hoagland's account of the mountain lion to Matthiessen's account of the African elephant. What do their discussions of the sightings, population, and terrain changes have in common? How is seeing wildlife a "part of privileged living" for Americans? Is the same true for Africans?

3. "It is not necessary to choose between being a country man and a city man, as it is to decide, for instance, some time along in one's thirties, whether one is an Easterner or a Westerner." With this statement Hoagland challenges one us/them opposition while emphasizing another. What aspects of the Easterner/Westerner rivalry are you familiar with? What might one gain by living in both the country and the city? How might this be preferable to inhabiting only the fringes

of settled regions? How can living only in a privileged location make it easier to ignore social issues linked to conservation?

WILLIAM KITTREDGE (b. 1932)

OWNING IT ALL (pp. 706–18)

1. Kittredge confesses to all the ways he and others "educated to believe in a grand bad factory-land notion" hurt the ecosystem of Warner Valley. What did they do, and what were the results?

2. How is Kittredge's account of being misled by the myth of the American West like Mowat's account of the wolf myth? How is Kittredge's statement—"only after re-imagining our myths can we coherently remodel our laws, and hope to keep our society in a realistic relationship to what is actual"—true for both his and Mowat's experiences?

3. We have read many writers who use war metaphors to refer to human relations with the land and wild animals. Kittredge, Snyder, Johnson, and others have directly linked our nation's military activities and attitude to how we interact combatively with the natural world. Snyder relates clearcutting to mowing down the very old and very young in Vietnam. Kittredge connects our fascination with agricultural technology to "the dream of power over nature and men, which I had begun to inhabit while playing those long ago games of war." Do you see any such connections in the world today? What role did the land and natural resources play in the Persian Gulf War? Are violent, political actions toward people and land still rationalized as "doing God's labor and creating a good place on earth"?

WENDELL BERRY (b. 1934)

AN ENTRANCE TO THE WOODS
and THE MAKING OF A MARGINAL FARM (pp. 718–36)

1. Understanding one's "destructive history," explains Berry, has a "paralyzing effect unless [you find] healing work." What kind of healing work does Berry suggest? What other such labor can you think of? Berry tells us, "[A] man cannot despair if he can imagine a better life,

and if he can enact something of its possibility." What is the importance of marrying action to imagination?

2. Compare "An Entrance to the Woods" with Thoreau's "Ktaadn." How does each man react to the landscape he is entering? What kind of tension do they feel between a desire for human community as buffer against the indifferent wilderness and a desire for solitude and what it provides? How does Berry's modern day discussion of wilderness differ from Thoreau's? How is it similar?

3. In these two essays, Berry describes a land of deciduous woods, rolling Appalachian mountains and hills, deep and lush river valleys—one of the "closed" and "easy" places that contrast with western landscapes. Whether certain types of people are drawn to certain types of landscapes, or whether landscapes shape people in unique ways—or both—consider how Berry's writing differs from the works of writers from other regions. Look at tone, style, syntax, and so on. Are there certain general characteristics you find in writers from the same landscape?

N. SCOTT MOMADAY (b. 1934)

THE WAY TO RAINY MOUNTAIN (pp. 737–42)

1. What does Momaday mean when he says his grandmother "bore a vision of deicide"? How did killing the buffalo and taking the land ultimately result in deicide?

2. How does Momaday's story demonstrate the type of upbringing that Standing Bear described in his essay? Though they have similar perspectives and goals, each chooses a different way to present his message. What form does each choose, and how does the form affect your response?

3. Momaday uses the same symbolic language to discuss his grandmother and the land. Does this seem strange? What effect does this have on Momaday's tone and his perceptions? How is his account of elders and generational knowledge like Snyder's and Berry's accounts? Have you had someone you consider an elder in your life? What wisdom and connections did they pass along to you?

SUE HUBBELL (b. 1935)

From A COUNTRY YEAR (pp. 743–49)

1. Why does Hubbell believe that buying meat from the grocery store is less responsible than killing it herself?

2. How does Hubbell go about trying to find a renewed sense of identity, to "be awake and aware . . . to accept vulnerability and death," as Merton says? How does she create a structure for both living and knowing the land?

3. Hubbell refers to the difference between the long-time locals who "have lived off the land from necessity" and the transplanted intellectuals who come from ravaged places with high ideals and little practical knowledge. This insider/outsider tension often exists within grassroots environmental organizations. What difficulties might this tension pose for groups trying to agree on ideology, purpose, and action? How can diversity within environmental groups be an advantage as well?

TIM ROBINSON (b. 1935)

TIMESCAPE WITH SIGNPOST (pp. 750–54)

1. What is the significance of Robinson's dolphin encounter? Why does he describe it as an "instance of wholeness beyond happiness"?

2. When Robinson writes of finding "some way of contributing to this society and surviving financially" on Aran, he echoes Berry's emphasis on making oneself responsibly at home. How does Robinson make himself responsibly at home? How do Berry's and Robinson's circumstances and experiences differ?

3. Robinson dreams of writing "the guide-book to the ultimate step" but knows it is impossible. He equates the complexities of human awareness—all the layers—to the actual strata we walk on with its history, biology, geology, and so on. What ideals does "the ultimate step" represent? How have other nature writers approached the struggle between what cannot be said in language but trying anyway?

CHET RAYMO (b. 1936)

THE SILENCE (pp. 754–59)

1. What extended metaphor does Raymo use? What does it help him do?

2. Wallace Stegner considers silence a threatened natural resource. According to Raymo, the universe is full of silence. Why is silence important to both writers? What does it inspire in us and teach us?

3. Raymo argues that silence is necessary for language. What do you think about the relationship between silence and communication? If all that depends upon verbal communication—writing, singing, talking, naming—depends upon silence, what will happen as our world becomes more and more noisy? Does it matter what type of noise fills the silence?

JIM HARRISON (b. 1937)

THE BEGINNER'S MIND (pp. 759-66)

1. Not only does Harrison go camping with Doug Peacock, but he also knows Gary Snyder and Gary Nabhan—writers in this anthology. Why might their purpose and subject matter encourage a close rather than competitive community among nature writers? Use examples from Harrison's essay to support your opinion.

2. Harrison's tone is reminiscent of Abbey's. How are their opinions and literary style similar?

3. Harrison wrote this essay for a collection of nonfiction the Nature Conservancy published, *Heart of the Land: Essays on Last Great Places*, to help raise money for conservation and spread the word. (This explains Harrison's first line.) The Nature Conservancy works to preserve land as private holdings, as opposed to public lands managed by state or federal government. Find out the different levels of protection offered by our three main public lands agencies: the National Forest Service, the National Park Service, and the Bureau of Land Management. (All these agencies' websites, as well as that of the National Wilderness Preservation System, are good places to

look.) What are the pros and cons of preserving land as private holdings? As public holdings? Which do you feel is preferable and why?

FREEMAN HOUSE (b. 1937)

IN SALMON'S WATER (pp. 766–72)

1. *Coevolution* is a term used to describe the shared evolution and mutually complementary adaptation of two or more species. How does House describe the mutual adaptation of humans and salmon on the northwest coast? How has their coevolution been disturbed?

2. Compare House's account of catching the salmon as a part of restoration efforts to McPhee's account of the yearly sedating and weighing of black bears. How is wildlife management complicated when a species has been or is a bioregion's primary food source (for humans as well as other creatures)?

3. Many nature writers have observed how much more valued a creature or place becomes when it becomes rare. "Our sense of relationship is replaced by fear of scarcity," claims House. Think about your approach to food. Do you experience it as a relationship or an entitlement? Have you ever killed your own meat, planted a garden, stood in food rationing lines? If you could eat only what was produced in your bioregion, what would that be?

WILLIAM LEAST HEAT-MOON (b. 1939)

UNDER OLD NELL'S SKIRT
and ATOP THE MOUND (pp. 773–81)

1. In both these selections Heat-Moon pays special attention to names and naming. Why does the preacher in "Under Old Nell's Skirt" object to naming things within nature?

2. In "Atop the Mound," Heat-Moon says he has a "woodland sense of scale and time." How does walking toward a specific goal across open ground affect his experience of time and space? How does his description of hiking in woods differ from Berry's in "An Entrance to the Woods"? What might account for their different impressions?

3. Heat-Moon claims names bring a sense of connection. Some would claim, however, that naming nature encourages sentimentalism or idolatry. Others say it encourages a sense of ownership or mastery over nature. Still others say it makes it possible to distinguish species—which, they say, is the only way we will be able to recognize when certain ones start to disappear. What role do you think naming plays in our culture and in your life? Why do we name sports teams after animals? Why do you think we name hurricanes and tropical storms and not tornadoes or earthquakes?

BRUCE CHATWIN (1940–1989)

From THE SONGLINES (pp. 781–87)

1. Chatwin says he was told that Aboriginals believe the earth gave humans language. How is this belief demonstrated in Aboriginal cultures?

2. From what Chatwin explains about the Songlines, how is Momaday's visit to his ancestor's place and Devil's Tower a walkabout in which he sings up his people's land?

3. In "Songlines" Chatwin makes many connections and comparisons between western culture and Aboriginal culture. Through what practices do you express your beliefs and values? What are the rituals of your work and play? What systems do you use to mark out territory and/or organize your life? How do you assert your right to exist and your unity with the world?

MAXINE HONG KINGSTON (b. 1940)

A CITY PERSON ENCOUNTERING NATURE
(pp. 787–90)

1. How do Kingston's imagery and language demonstrate the fact that she is a city person?

2. Hoagland stated that one needn't choose between being a city or country person. How does Kingston's essay support Hoagland's claim? What types of things does Kingston notice? How does she

show her interest in the intersection between natural and manmade things?

3. Kingston says, "[A] new climate helps me to see nature." Describe seeing an unusual plant, animal, or landform. Was it in a familiar or a new environment? Do you think you are more alert and perceptive in new environments?

JOHN HANSON MITCHELL (b. 1940)

From LIVING AT THE END OF TIME (pp. 790–96)

1. Why did Mitchell once feel that the North American landscape was "incomplete or lonely"?

2. Mitchell layers many personal and cultural memories on his experience of the northeastern landscape. How do these associations add to his sense that it is a "haunted land"? Mitchell seems to agree with Kingston that new places help one to see nature when he states that perhaps Thoreau left Walden Pond because "he realized that staying on would have dulled the experience." Does Mitchell contradict himself? Explain.

3. Why does Mitchell give us the story of Thoreau's illness? How did this famous nature writer engage with his own death? Did he act as if he were "a part and parcel of nature"? What surprised, disappointed, or impressed you about the way Thoreau met death? Why?

RICHARD K. NELSON (b. 1941)

THE GIFTS (pp. 797–810)

1. How does Nelson reveal his absorption of the Koyukon ways while also showing us how he is shaped by western civilization? How does this affect his tone and credibility?

2. Compare Nelson's worry over winter and efforts to get his family's food supply in with LeSueur's description of the canning and preserving that people did on the Great Plains in order to survive. What are Nelson's reasons for hunting? Do reasons and methods make a difference in the ethics of hunting? Think of Hubbell's regret that she

can't take responsibility for killing her own meat. How does hunting draw Nelson deeper into the land community?

3. Nelson's stories show how much the hunter and the watcher have in common. They each share a reliance on luck and both relationships "are founded in the same principles . . . the same reciprocity." How does Nelson emphasize the communing with nature over the conquering? How does his interaction with the island community make Nelson more aware of himself as an animal?

JOSEPH BRUCHAC (b. 1942)

THE CIRCLE IS THE WAY TO SEE (pp. 811–18)

1. For Bruchac, and many native cultures with oral traditions, stories live and walk around. Are stories valued as much in western culture? What is their role in our society?

2. Both Bruchac and Nelson write about hunting. What beliefs about the appropriate behavior of hunters do they share? What else do they have in common?

3. Bruchac writes of the "lesson stories" Native Americans were given "to remind us of our proper place." Every culture has lesson stories: the Greek, Roman, and Norse myths; Judeo-Christian parables; European fables and fairy tales; even family stories passed down through the generations. Think of the lesson stories you were raised with. What were they supposed to remind you of? Were they concerned with more than human interactions?

FRANKLIN BURROUGHS (b. 1942)

OF MOOSE AND A MOOSE HUNTER (pp. 819–32)

1. How does Burroughs counter the moose image we see in catalogs, products, and media? What traits does he associate with the moose?

2. Compare Burroughs's description of the controversy over moose hunting with Hubbell's account of the folks fighting the dam. Why do newcomers and natives tend to take different approaches to these issues? How does Burroughs bring out these tensions in his essay?

3. What is Burroughs referring to when he talks about "the Arcadian daydream of man and nature harmoniously oblivious to the facts of man and nature"? How do Bruchac's lesson stories offer an alternative to this Arcadian daydream? Is it one our culture could adopt in "the context of our historical violence," as Burroughs says?

DOUG PEACOCK (b. 1942)

THE BIG SNOW (pp. 832–41)

1. Why did Peacock seek out grizzlies and wilderness when he returned from Vietnam? With what did the ritual of watching the grizzlies provide him?

2. Compare Peacock to Nelson. What led each man to spend so much time in the wilderness? What roles do violence, humility, practicality, awareness, and spirituality play in their lives? What rituals or ceremonies do each practice?

3. Peacock is an angry, anti-social man who has been trained to kill and has seen unthinkable atrocities. How does his personality contrast with those of other nature writers? It has been said that war makes men act like beasts. Why is Peacock's antidote to the pain and madness of Vietnam to become more of an animal himself?

ROBERT FINCH (b. 1943)

DEATH OF A HORNET (pp. 841–44)

1. How did the hornet become entangled in the spider web? How does the cause of this entanglement influence the story?

2. Compare Finch's account of how the spider kills and prepares the hornet to Nelson's account of killing and dressing the deer. Nelson describes the process as "a ritual of respect." Is the same true for the spider? Is the spider the opposite of a hunter?

3. "Defense in insects, as with us, seems to be founded not on the ability to survive but on the resolution to keep from forgiving as long as possible." What do you think of Finch's statement? Think about other prey you have read about, such as the zebra Matthiessen

describes as accepting the wild dog's attack. Are human defense mechanisms tied more to pride than to survival? What connection is there between Finch's statement and Bruchac's observation that "human self-importance is a big part of the problem"?

LINDA HASSELSTROM (b. 1943)

NIGHTHAWKS FLY IN THUNDERSTORMS
(pp. 845–49)

1. In what ways have nighthawks taught Hasselstrom about the Great Plains? What do they symbolize for her?

2. How was Hasselstrom's ranch life similar to and different from Berry's life on a farm?

3. Hasselstrom's father's motto was "What can happen, will." How is this type of acceptance a survival instinct? Why do those who work closely with the land have this sense when many people working with technology do not?

TRUDY DITTMAR (b. 1944)

MOOSE (pp. 850–62)

1. Dittmar's essay focuses on several different things. Why is her discussion of the moose necessary to her discussion of evolution and altruism? What rhetorical techniques does she use to interweave these themes?

2. Compare Dittmar's description of moose in the Wyoming high country to Burroughs's impression of moose in Vermont. How are these descriptions alike? What contributes to their differences?

3. Dittmar uses the moose to help her understand how to live with "the beauty of the cold world," where kindness and compassion seem not to come from goodness but from genetic efforts to survive. What are the consequences of the socio-biological view that all altruism is ultimately selfish? How can we "also have a big share of free will"?

ALICE WALKER (b. 1944)

AM I BLUE? (pp. 863–67)

1. What connections does Walker make between domestic animals and human slavery?

2. Walker is shocked to realize she had "forgotten that human animals and nonhuman animals can communicate quite well." How are her views on interspecies communication like and unlike Kumin's?

3. Kumin says that women tend to be more sympathetic to and understanding of horses. Walker compares the plight of domestic horses to the plight of women, slaves, and minorities. Why do language and communication play such an important role in the power relationships she critiques?

ANNIE DILLARD (b. 1945)

HEAVEN AND EARTH IN JEST, LIVING LIKE WEASELS, *and* TOTAL ECLIPSE (pp. 867–91)

1. In "Heaven and Earth in Jest," what is Dillard's reason for keeping this kind of journal? What paradoxes does she explore?

2. Compare Dillard's tone to Bass's. How does each establish credibility? How are their purposes different, and what effect does that difference have on their essays?

3. Dillard's writing focuses on how to live in the terrifying and beautiful world. After reading these three selections, how would you answer her question in "Heaven and Earth in Jest": "What do we think of the created universe, spanning an unthinkable void with an unthinkable profusion of forms?"

JAN ZITA GROVER (b. 1945)

CUTOVER (pp. 891–99)

1. Why does Grover include the story of Perry's leg in this essay?

2. The Transcendentalists read nature as a text to be decoded into spiritual meaning. Why does Grover resist this "theory of correspondence"? She claims to be "suspicious of metaphor," but does she avoid metaphor completely? How does her approach different from that taken by the Transcendentalists such as Emerson, Thoreau, and Whitman?

3. "Which forest am I mourning?" asks Grover, scrutinizing the common "sentimental response" to altered landscapes. How does her account of succession cycles complicate our ideas of what a diminished landscape should look like? Grover suggests that bodies are another type of landscape and that they reveal their disturbance histories as much as the earth. What connections between bodies, land, and cycles are at play in the question "Is there a sense in which this leg can be viewed as a creation instead of an annihilation?"

BARRY LOPEZ (b. 1945)

From ARCTIC DREAMS and THE AMERICAN GEOGRAPHIES (pp. 900–923)

1. What is the significance of the Eskimo's statement in "Migration": "We do not believe. We fear"? How is this similar to Hasselstrom's father's motto: "what can happen, will"?

2. In "Lancaster Sound," Lopez says, "[I]nterpretation can quickly get beyond a scientist's control. When asked to assess the meaning of a biological event . . . they cannot say what it means, and they are suspicious of those who say they know." Compare this suspicion with Grover's resistance to metaphors that explain what a landscape means. In "Ice and Light" Lopez says that we bring our own "metaphorical tools of the mind" to bear on landscapes and that "it is hard to imagine that we could do otherwise." However, he warns against "finding our final authority in the metaphors rather than in

the land." How does Lopez explore what he sees as a necessary balance of awareness and experience, landscape and memory?

3. Like Snyder, Lopez laments in "American Geographies" how little our lives are informed by local knowledge. "In forty thousand years of human history," Lopez claims, "it has only been in the last few hundred years or so that a people could afford to ignore their local geographies as completely as we do and still survive." In what ways are you cut off from your local geography? Do you know where your water comes from and where it goes? Are the interactions between the plants and animals familiar to you? What landforms affect your climate? What people inhabited this place before your people? Do you fear anything related to your geography? Do you agree with Lopez that "the geographies of North America . . . are threatened—by ignorance of what makes them unique, by utilitarian attitudes, by failure to include them in the moral universe, and by brutal disregard"?

SCOTT RUSSELL SANDERS (b. 1945)

BUCKEYE (pp. 924–29)

1. Though Sanders learned to recognize different trees when he was a boy, why did he not know their scientific or book names till later?

2. How does Sanders's account of the land his father passed down to him demonstrate the generational or local knowledge that Snyder and Lopez say is so important? Why is this knowledge—in Sanders's family and in general—disappearing?

3. "The Ohio landscape never showed up on postcards or posters," remarks Sanders. What images do show up on postcards and advertisements? What composite concept of a general American geography—like the one Lopez discusses—do these images create? What is the message behind the generalization?

DAVID RAINS WALLACE (b. 1945)

THE HUMAN ELEMENT (pp. 930–36)

1. According to Wallace, what is the difference between humans and other organisms?

2. Wallace claims that in addition to intellectual knowledge there is "emotional knowledge." How do the following quotations from other nature writers in this collection rely on emotional knowledge? Why is this type of knowledge important?

> [Wolves] don't have thumbs. All they've got is teeth, long legs, and—I have to say this—great hearts.　　　—RICK BASS

> I tell you I've been in that weasel's brain for sixty seconds, and he was in mine.　　　—ANNIE DILLARD

> It was a red-tailed hawk for sure; and it was my father. Not a symbol of my father, not a reminder, not a ghost, but the man himself, right there circling in the air above me. I knew this as clearly as I knew the sun burned in the sky.　　—SCOTT RUSSELL SANDERS

> Never mind the issue of knowing; we should assume that the power is here and let ourselves be moved by it.　　—RICHARD NELSON

> The answer must be, I think, that beauty and grace are performed whether or not we will or sense them. The least we can do is try to be there.　　　—ANNIE DILLARD

3. We usually think of evolution as science, and science as the opposite of myth. In what ways does Wallace show that western culture interprets events and facts into myths that separate us from the rest of nature? Wallace advocates conscious cultural evolution in which our myths and, therefore, our behavior change. What types of new myths does he think we need? What new myths do you think we need?

ALISON HAWTHORNE DEMING
(b. 1946)

WOLF, EAGLE, BEAR: AN ALASKA NOTEBOOK
(PP. 937–43)

1. What tribe, according to Deming, is she from? How does her writing reflect her membership in this tribe as opposed to the Koyukon or Tlingit people of Alaska?

2. Deming shares Wallace's awareness of how evolution connects all organisms and how free will has allowed our species to evolve more rapidly than mere biology would allow. Compare their views on the animal nature of humans, especially survival instincts and fear. How is the grizzly to which Deming "left the high woods" like the giants Wallace discusses?

3. Deming emphasizes her perspective as a woman and examines how women and men in different cultures interact with grizzlies. She argues that in western culture women and men have a different relationship to fear. What evidence does she give? How do you feel about her statements comparing men and women?

GRETEL EHRLICH (b. 1946)

FRIENDS, FOES, AND WORKING ANIMALS (pp. 944–49)

1. What examples of outsiders' "patronizing attitude toward animals" does Ehrlich give?

2. For Ehrlich, caring for domestic animals such as sheep, cattle, and dogs is as much a part of interacting with the land as encountering bighorn sheep, bobcats, and elk. "[W]e give ourselves as wholly to the sacrament of nurturing as to the communion of eating their flesh," she writes. How is Ehrlich's attitude toward animals like and unlike Nelson's?

3. Ehrlich blurs many of the distinctions between wild and domestic animals. How do you feel about these categories? Ehrlich's view is the opposite of Dillard's when she comments that steers are "a human product like rayon. . . . You can't see through to their brains

as you can with other animals." What do you think? Is it easier to make clear distinctions between wild and nonwild animals than between wilderness and nonwilderness?

ELLEN MELOY (b. 1946)

THE FLORA AND FAUNA OF LAS VEGAS (pp. 950–59)

1. Meloy explains how people get water from rivers, snowmelt, and aquifers. What are the differences between these sources?

2. How is Meloy's tone reminiscent of Abbey's? Compare their use of sarcasm. Is it equally effective in both authors' work? What does sarcasm accomplish?

3. It might be said that all writers are naturalists of their own species. Meloy turns the eye of a naturalist on people in Las Vegas. How does she use her observations to explore larger issues such as the relationship between the real and artificial? Is *real* the same as *natural*?

EMILY HIESTAND (b. 1947)

ZIP-A-DEE-DO-DAH (pp. 959–66)

1. What qualities does Hiestand ascribe to the blue jay? How does she make sure that we will see the blue jay in a favorable light?

2. Hiestand describes birds as either specialists or generalists, with differing chances of survival in a world where habitat is rapidly being altered by humans. According to Baker, Bass, Pyle, Pollan, and others, what species are adapting? Are they all generalists, or are some specialist species making a comeback?

3. Just as Krutch did not want to hold other species to humans' moral standards, so Hiestand will not condone or condemn blue jays for eating other birds' eggs. More than a century after Knapp, Wilson, and others lamented human prejudices toward certain animals, Hiestand, Krutch, and others show how such prejudice continues. "I start to wonder," writes Hiestand, "if we Americans have decided to deny that class exists in our society mostly so that we can have the fun of unconsciously projecting it onto flora and fauna." What do you think

about this idea? Is there any way in which you consider plants and animals in terms of class? What might be the ramifications of making such projections in the twenty-first century?

LINDA HOGAN (b. 1947)

THE BATS (pp. 966–71)

1. How does Hogan describe a bat's relationship to language? Cite examples from her text.

2. How is Hogan's exploration of the nature of bat consciousness similar to what Wallace urges us to do? Unlike Dillard, who says she has been in a weasel's brain, and Williamson, who tells stories from the animals' perspective, Hogan states, "I see them through human eyes. . . . I don't hear the high-pitched language of their living, don't know if they have sorrow or if they tell stories . . ." How do these different approaches affect your response to the writer's message? Why is point of view so important in these essays and stories?

3. Were you surprised to read how bats were experimented on for war activities? Marine mammals are still being used for oceanic war and defense tasks. Compare our culture's relationship with bats to that of the southern tribes Hogan mentions. How does each culture look to bats for assistance? Why does Hogan suggest that bats, since "they live in double worlds of many kinds," can help us now? Why does she associate them with mercy?

ROBERT MICHAEL PYLE (b. 1947)

AND THE COYOTES WILL LIFT A LEG (pp. 971–99)

1. How do Pyle's light tone and wordplay work with and against his subject?

2. Both Pyle and Dillard talk about the importance of making an effort to encounter beauty and grace in our lives. What effect do their opinions concerning organized religion and God have on their writing? Does Pyle's activist approach lead him to use different rhetorical strategies than Dillard, with her contemplative approach?

3. Pyle writes, "[T]he present and near future could get downright depressing if it weren't for nature." Why does he mention only the *near* future? How does Pyle's view that nature is "the whole to which our greater allegiance belongs" influence her pessimism toward the next century and optimism toward the long term?

DIANE ACKERMAN (b. 1948)

WHY LEAVES TURN COLOR IN THE FALL (pp. 979–83)

1. How does Ackerman weave together the lyrical and more strictly informative parts of her essay? What affect does this approach have on your experience of the essay as a whole?

2. How is Ackerman's style similar to that of writers like Lewis Thomas and Chet Raymo? What techniques do these science writers use to explain complicated phenomena and keep the reader's interest?

3. Ackerman describes the experience of fall in an area thick with deciduous trees. How does she give you a sense of what it is like? "Leaves have always hidden our awkward secrets," she says. What do you think Ackerman means by this? Do falling leaves reveal pleasant details as well? Compare the landscapes Ackerman describes to more austere or less changeable ones such as the desert, plains, or high mountains. Which do you find more aesthetically pleasing and why?

JOHN DANIEL (b. 1948)

A WORD IN FAVOR OF ROOTLESSNESS (pp. 983–90)

1. What are the dangers of being boomers and nesters, according to Daniel?

2. How do Daniel and Peacock address the issue of territoriality—both in humans and in other creatures? How does each make a case for the wanderer?

3. Daniel points out that environmentalists can be self-righteous and intolerant and acknowledges that this attitude makes it harder for people to hear their message. For this reason, the rhetoric a nature writer chooses is extremely important. Have you found any of the

styles in this anthology to be overly preachy or judgmental? Which have been the most persuasive? Do you think Daniel follows his own advice?

DAVID QUAMMEN (b. 1948)

STRAWBERRIES UNDER ICE (pp. 990–1002)

1. According to Quammen, what is the difference between a glacier and an ice sheet? What lies above and below a glacier's equilibrium line?

2. Quammen uses metaphor to explain scientific facts and phenomena. How does he avoid, as Barry Lopez says, "finding . . . final authority in the metaphors rather than the land"?

3. What connections does Quammen make between events in his life, places he has been, and the scientific knowledge concerning different types of ice? How does this approach help readers understand the scientific concepts he presents? How does he use these concepts as metaphors—and for what reason?

LESLIE MARMON SILKO (b. 1948)

LANDSCAPE, HISTORY, AND THE PUEBLO IMAGINATION (pp. 1003–14)

1. What is Silko's objection to the term *landscape*?

2. Joseph Wood Krutch notes how the people of Western cultures have recoiled from our own waste, and Lewis Thomas points out the way Western culture has difficulty accepting death. How is Silko's account of the Pueblo Indian views different?

3. Silko credits her people's oral tradition with passing on the generational knowledge necessary to their living in place. How do stories act like maps in the Pueblo culture? What stories act like maps in your culture?

JAMAICA KINCAID (b. 1949)

ALIEN SOIL (pp. 1015–22)

1. According to Kincaid, how are class and wealth manifested in the Antiguan landscape?

2. Compare Kincaid's discussion of Thomas Jefferson and American ideals with Bruchac's account of the model on which the American forefathers based the early democracy. What ironies and complexities do they reveal?

3. Kincaid highlights three approaches to plant cultivation: small, subsistence gardens; ornate, decorative gardens; and the raising (or just harvesting) of large-scale crops—including trees. "The study of plants is now called agriculture," she writes. If botany is about "awe and poetry" and small-scale agriculture is about feeding a family or community, what is large-scale agriculture about? How is the demise of the family farm in America like and unlike what has happened to Antigua?

DAVID JAMES DUNCAN (b. 1952)

NORTHWEST PASSAGE (pp. 1022–27)

1. How and why does Duncan establish the differences between him and Jered?

2. Compare Duncan's encounter with the coho to Dittmar's encounter with the moose. What does each writer learn from these creatures? How does each create the tension in their accounts?

3. Duncan describes his adolescence as a time when he longed to escape to better or more exotic places. What does watching the salmon teach him about escape? Think of Daniel's essay on rootlessness. How do recognize an urge to escape— responsibilities as well as places—and what characterizes an attraction to our "true place?"

RAY GONZALEZ (b. 1952)

THE THIRD EYE OF THE LIZARD (pp. 1028–33)

1. According to Gonzalez, what is the purpose of the lizard's third eye?

2. Compare Gonzalez's essay to Nelson's. How is hunting portrayed? Does Gonzalez find a communal relation to his prey? What difference does it make that one hunts mammals and one hunts reptiles?

3. Like Snyder, Gonzalez uses his own poetry within his essay. What advantage does quoting poetry have over paraphrasing it? What qualities does poetry have that distinguishes it from prose? Of all the prose writers you have read, whose language seems most wild and poetic? Why? Do you prefer the more poetic writers?

VICKI HEARNE (b. 1952)

CALLING ANIMALS BY NAME (pp. 1034–38)

1. What does Hearne feel is most responsible for distancing humans from other animals?

2. Maxine Kumin thinks that women make better animal trainers because they are more empathic, nurturing, and instinctive. Does Hearne's essay support or contradict Kumin's view?

3. Hearne distinguishes between names used to label and names used to invoke. She shows how names can distance one from things as well as connect and include other creatures in "the moral life." Think of times you have been referred to by a label. Are addresses such as "hey, handsome" or "little lady" a way of connecting or distancing? From Hearne's perspective, what kind of naming is the Linnaean binomial system?

GARY PAUL NABHAN (b. 1952)

From THE DESERT SMELLS LIKE RAIN . . . (pp. 1038–42)

1. What does Nabhan say are the likely factors that make "real rain" so much better for growing crops than simulated rain?

2. How does Nabhan's essay, like Meloy's discussion of Las Vegas water use, address the issues of water management? How does each demonstrate that fresh water is not an unlimited resource?

3. Why does Nabhan incorporate dialogue and a linguistic discussion of the Papago people's language into his essay? How does a group's language reflect its relationship to nature? How do common English idioms (for example, "I was treated like dirt") reflect our assumptions about nature and our alienation from it?

LOUISE ERDRICH (b. 1954)

BIG GRASS (pp. 1043–47)

1. What does the phrase "all flesh is grass" mean to Erdrich?

2. In this essay, Erdrich reads an aspect of nature like a text when she explores metaphorical interpretations of grass. Other nature writers, especially Grover, have cautioned against this approach. What is your reaction to it?

3. Erdrich says that buffalo "always made sense." What is the relationship between buffalo and native big grass? What has each been replaced with in certain areas? How are their presence and absence linked?

DAVID MAS MASUMOTO (b. 1954)

PLANTING SEEDS (pp. 1047–51)

1. Why did Masumoto first start planting cover crops again?

2. Look up the terms "agribusiness" and "monoculture" if you are not familiar with them. How do these trends threaten family farms like Masumoto's and Berry's?

3. Masumoto makes connections between raising a child and planting seeds. What do discussions about good soil, hands-on interaction with the land and crops, and knowing the difference between what is best for the crop and what is easiest for the farmer have in common with discussions of good and bad methods of raising children?

SHARMAN APT RUSSELL (b. 1954)

GILA WILDERNESS (pp. 1051–62)

1. Russell refers to the Wilderness Act of 1964 many times. According to Russell, the Wilderness Act, and Leopold (whom Russell discusses), why should we designate and preserve wilderness areas?

2. Both Russell and Deming see differences between how and why men and women go into the wilderness. What differences do they identify? Do they agree?

3. One result of the personification of the wilderness as mother, virgin, and exotic lover is that it has been portrayed as a place to which men, not women, go. How has this macho image shaped our understanding of the relationship between wilderness and solitude? Why does Russell describe *all* the human interaction on her trip, especially that between the women and children?

EVELYN WHITE (b. 1954)

BLACK WOMEN AND THE WILDERNESS (pp. 1062–67)

1. What is White afraid will happen if she ventures outside?

2. Thoreau writes in "Walking": "How near to good is what is *wild!*" White's essay highlights the conflicting meanings acquired by the words *wild* and *wilderness*—free and exciting versus uncontrolled and threatening, a place of idealized, pristine nature versus a dangerous place full of lawless creatures (including humans). What differences in White's and Thoreau's cultural positions led to their radically different responses toward the outdoors?

3. Are there outdoor places where you have felt unsafe? Describe such

a place and the circumstances of your being there. Analyze how your gender, class, ethnicity, and any other aspect of your cultural position may have affected your feeling of safety or belonging. Try to recall what form the threat took in your mind. Was it visible or invisible? Looking back, do you feel you overreacted?

BARBARA KINGSOLVER (b. 1955)

HIGH TIDE IN TUCSON (pp. 1068–78)

1. Kingsolver divides her essay into four sections. Two are narratives or scenes, and two are introspective commentaries. How do these sections play off one another?

2. Kingsolver writes that "it's starting to look as if the most shameful tradition of western civilization is our need to deny we are animals." This echoes Wallace Stegner, who states in "Coda: Wilderness Letter" that "one means of sanity is to retain a hold on the natural world, to remain, insofar as we can, good animals." Kingsolver highlights the habits and drives we share with other creatures. Which of these are causing us problems now?

3. Like Daniel, Kingsolver sees moving to nonnative places as a complex act of relinquishing old ties and acquiring new allegiances. How does each feel about American culture's emphasis on mobility and independence? How does Kingsolver expand the notion of the importance of belonging somewhere by advocating "renewing our membership in the animal kingdom"?

MICHAEL POLLAN (b. 1955)

WEEDS ARE US (pp. 1078–90)

1. Pollan points out that there are as many different definitions of the term *weed* as there are of the word *wilderness* and he notes how often it is used metaphorically. What specific definitions of *weed* does he consider? Which does he settle on?

2. Like Grover, Pollan calls our attention to altered landscapes and to

our assumptions about what their pure, natural state must have been. How do Pollan's discussion of weeds and Grover's account of succession cycles complicate our understanding of landscapes?

3. Pyle praises the weed's adaptability, saying, "[A]ll life in the ravaged land is a bunch of weeds—survivors, coping and adapting under adversity." Likewise Pollan equates weeds with surviving in an altered land, but also implicates them in damaging the land. If, as Harland says (when Pollan quotes him), "[M]an is by definition the first and primary weed under whose influence all other weeds have evolved," is our weedlike adaptability something to be proud of? Or is it a characteristic that implicates us in degrading the land? After reading Pollan, do you associate weeds with wildness or cultivation?

TERRY TEMPEST WILLIAMS (b. 1955)

THE CLAN OF ONE-BREASTED WOMEN (pp. 1091–98)

1. What is the Clan of One-Breasted Women? What phenomena does Williams hold responsible for the membership of her family's women in this clan?

2. Like Ursula K. Le Guin, Williams writes from an ecofeminist perspective. How does each writer protest the patriarchal institutions and practices of war?

3. There was a time when Williams quit writing to work full-time as an activist. Rick Bass says, "You have to decide whether the fight requires art or advocacy." Do you think nature writing can be as effective as protests, letter-writing campaigns, or nonviolent civil disobedience?

JANE BROX (b. 1956)

BALDWINS (pp. 1098–1101)

1. According to Brox, what practices have changed in commercial apple farming? Why?

2. Nature writers repeatedly allude to the value of scarcity, how the worth of something increases as it becomes more rare. For instance, Michael Pollan points out that pure white flowers are in the botani-

cal upper class because they are the least common. How does this principle affect Baldwin apples? How does the "fear of scarcity," as House says, lead of the commodification of rare apples, unblemished vistas, endangered animals, and silence?

3. Brox stresses the irony of how the media uses our nostalgia for simpler times to manipulate us. "The ones who lived [the old life] can't afford the price of these goods," she points out. How does Brox use the Baldwins as a symbol of all we are losing and of our attempts to hold onto the past? Can you give other examples of scarce items being sold to the highest bidder?

DAVID ABRAM (b. 1957)

THE ECOLOGY OF MAGIC (pp. 1101–14)

1. How does Abram use scenes to convey his change in perspective?

2. Wallace describes how cultures rooted in myths projected "human consciousness onto nature." Abram discusses cultures rooted in animism. How does their treatment of nature differ from that of the cultures Wallace discusses? Compare Abram's and Wallace's discussions of nonhuman consciousness. Where do their ideas converge and diverge?

3. Abram contrasts the Judeo-Christian notions of heaven and hell with the sacred view of earthly cycles held by many people with animistic beliefs, such as the Balinese and Koyukon. What is the difference between the Balinese idea of magic and the Judeo-Christian idea of miracles? How does the belief that human bodies and consciousness cycle back into the earth, air, and other creatures affect human and nonhuman interaction among animistic cultures? How does this perspective add meaning to Leopold's phrase "thinking like a mountain"?

RICK BASS (b. 1958)

From THE NINEMILE WOLVES (pp. 1114–19)

1. What does Bass mean by "we're all following the wolf"?

2. Remember what you've read about human/wolf interaction in Audubon, Seton, and Leopold. What new episode in this history does Bass relate? What is his purpose in writing this piece?

3. Bass, trained as a petroleum geologist, says he has given up his "science badge" and can "say what I want to say." How does his dismissal of an objective stance affect his writing? In another essay, "The Blood Root of Art," Bass states: "You have to decide whether to use numbers or images: you have to decide whether the fight requires art or advocacy—and try to have an awareness of where the one crosses over into the other" (*The Book of Yaak*). How do you see these two impulses at work in "The Ninemile Wolves"? Compare Bass to other nature writers who also worked as scientists, such as Lewis Thomas, René Dubos, Aldo Leopold, and E. O. Wilson. How do they balance images and facts? Which ones combine art with advocacy? What approach is most effective?

BILL McKIBBEN (b. 1960)

From THE END OF NATURE (pp. 1120–30)

1. When McKibben says that the idea of nature is extinct, what does he mean?

2. Compare McKibben's view and style to Pyle's. Contrast their rhetorical approaches to similarly dark subject matter. Do they reach the same conclusions? What part does style play in your reaction to their conclusions?

3. Abram presents the world from an animistic perspective in which nature is not indifferent, in which it is both other and familial. McKibben describes the world from a western perspective in which there is no magic or spirit in the rain, in which the world will soon be full of genetically engineered creatures. In Abram's vision we are of nature; in McKibben's we have killed nature. Do you think that nature is now "a subset of human activity" or not?

JANISSE RAY (b. 1962)

BUILT BY FIRE *and* FOREST BELOVED

(pp. 1130–35)

1. In "Built by Fire" what does Ray say the longleaf pine does to survive the lightning fires?

2. How does Ray apply Leopold's advice to think "like a mountain" to the southeastern coastal plains? How is Ray's personification of the pines and lightning similar to and different from animal portraits, such as Seton's and Williamson's? How do you feel about her giving voice to nature this way? Is she being empathic? Arrogant? Open-minded? Imperialistic? Silly? Explain your reaction.

3. In "Forest Beloved," Ray invokes Bartram's and Muir's travels through the area. Why is it important for her to give the reader a sense of time and of change? How else does she do this? How does Ray weave together a relationship to the nonhuman world that is both Christian and animistic?